Water-Worked Bedload

Water-Worked Bedload:
Hydrodynamic and Mass Transport

Special Issue Editors

Paweł M. Rowiński
Subhasish Dey

MDPI • Basel • Beijing • Wuhan • Barcelona • Belgrade

MDPI

Special Issue Editors
Paweł M. Rowiński
Institute of Geophysics, Polish Academy of Sciences
Poland

Subhasish Dey
Indian Institute of Technology Kharagpur
India

Editorial Office
MDPI
St. Alban-Anlage 66
4052 Basel, Switzerland

This is a reprint of articles from the Special Issue published online in the open access journal *Water* (ISSN 2073-4441) from 2018 to 2019 (available at: https://www.mdpi.com/journal/water/special_issues/Water_Worked_Bedload)

For citation purposes, cite each article independently as indicated on the article page online and as indicated below:

LastName, A.A.; LastName, B.B.; LastName, C.C. Article Title. *Journal Name* **Year**, *Article Number*, Page Range.

ISBN 978-3-03921-301-6 (Pbk)
ISBN 978-3-03921-302-3 (PDF)

Cover image courtesy of Subhasish Dey.

Contents

About the Special Issue Editors vii

Paweł M. Rowiński and Subhasish Dey
Water-Worked Bedload: Hydrodynamics and Mass Transport
Reprinted from: *Water* **2019**, *11*, 1396, doi:10.3390/w11071396 1

Magdalena M. Mrokowska and Paweł M. Rowiński
Impact of Unsteady Flow Events on Bedload Transport: A Review of Laboratory Experiments
Reprinted from: *Water* **2019**, *11*, 907, doi:10.3390/w11050907 4

Ellora Padhi, Subhasish Dey, Venkappayya R. Desai, Nadia Penna and Roberto Gaudio
Water-Worked Gravel Bed: State-of-the-Art Review
Reprinted from: *Water* **2019**, *11*, 694, doi:10.3390/w11040694 20

Ronald Möws and Katinka Koll
Roughness Effect of Submerged Groyne Fields with Varying Length, Groyne Distance, and Groyne Types
Reprinted from: *Water* **2019**, *11*, 1253, doi:10.3390/w11061253 39

Li Zhang, Pengtao Wang, Wenhai Yang, Weiguang Zuo, Xinhong Gu and Xiaoxiao Yang
Geometric Characteristics of Spur Dike Scour under Clear-Water Scour Conditions
Reprinted from: *Water* **2018**, *10*, 680, doi:10.3390/w10060680 50

Li Zhang, Hao Wang, Xianqi Zhang, Bo Wang and Jian Chen
The 3-D Morphology Evolution of Spur Dike Scour under Clear-Water Scour Conditions
Reprinted from: *Water* **2018**, *10*, 1583, doi:10.3390/w10111583 63

Federica Antico, Ana M. Ricardo and Rui M. L. Ferreira
The Logarithmic Law of the Wall in Flows over Mobile Lattice-Arranged Granular Beds
Reprinted from: *Water* **2019**, *11*, 1166, doi:10.3390/w11061166 78

Artur Radecki-Pawlik, Piotr Kuboń, Bartosz Radecki-Pawlik and Karol Plesiński
Bed-Load Transport in Two Different-Sized Mountain Catchments: Mlynne and Lososina Streams, Polish Carpathians
Reprinted from: *Water* **2019**, *11*, 272, doi:10.3390/w11020272 93

Lei Huang, Hongwei Fang, Ke Ni, Wenjun Yang, Weihua Zhao, Guojian He, Yong Han and Xiaocui Li
Distribution and Potential Risk of Heavy Metals in Sediments of the Three Gorges Reservoir: The Relationship to Environmental Variables
Reprinted from: *Water* **2018**, *10*, 1840, doi:10.3390/w10121840 108

Łukasz Przyborowski, Anna Maria Łoboda and Robert Józef Bialik
Experimental Investigations of Interactions between Sand Wave Movements, Flow Structure, and Individual Aquatic Plants in Natural Rivers: A Case Study of *Potamogeton Pectinatus* L.
Reprinted from: *Water* **2018**, *10*, 1166, doi:10.3390/w10091166 125

About the Special Issue Editors

Paweł M. Rowiński, Professor, Member of the Polish Academy of Sciences. In 2015, he was elected as the Vice President of the Polish Academy of Sciences (in 2019 he has started his second term). From 2008 to 2015 he was the CEO of the Institute of Geophysics, Polish Academy of Sciences, and earlier (2004–2008), the Research Director of that Institute. He was also the Co-Founder of the Earth and Planetary Research Centre (GeoPlanet). During 2009–2015, he served as the first Chairman of the Board of Directors of the GeoPlanet. In 2018, he was elected as the Vice Chair of the International Association for Hydro-Environment Engineering and Research *IAHR* Europe Division Leadership Team. In 2018, he was elected a member of Board of ALLEA—the European Federation of Academies of Sciences and Humanities. His research areas encompass mathematical modeling of hydrological processes, fluvial hydraulics, river turbulence, pollution, heat, and sediment transport in rivers, two-phase flows, chaotic dynamics, water balance in a catchment, flow–biota interactions, expert systems, neural networks, adaptive environmental assessment, management; he has also made important contributions to experimental hydraulics He has more than 160 scientific publications to his credit. He has been a co-author and/or co-editor of 18 scientific volumes (including this one). He was awarded a number of recognized prizes, among them, the stipend of the Foundation for Polish Science for outstanding young scientists and scholarship of the Central European University. In 2015, he was awarded Bene Merito distinction by the Polish Minister of Foreign Affairs as an acknowledgment of his services that contribute to strengthening Poland's status in the international arena. He was an Associate Editor of the *Hydrological Sciences Journal* (IAHS Press, Wallingford, UK) and is currently the Editor-in-Chief of the Monographic Series *Geoplanet: Earth* and *Planetary Sciences*, Springer Verlag.

Subhasish Dey, Professor, Department of Civil Engineering, Indian Institute of Technology Kharagpur, West Bengal 721302, India. He is a hydraulician and educator. He is known for his research on the hydrodynamics throughout the world and acclaimed for his contributions to developing theories and solution methodologies of various problems in applied hydrodynamics, turbulence, boundary layer, shallow fluid flows, and sediment transport. He also holds an adjunct professor position in the Physics and Applied Mathematics Unit at Indian Statistical Institute Kolkata (2014–2019) and a Distinguished Visiting Professor of Tsinghua University position in the Department of Hydraulic Engineering, Tsinghua University, Beijing, China (2016–2019). He is an author of the textbook Fluvial Hydrodynamics, published by Springer. He has published over 175 research papers in refereed journals. He is an associate editor of the *Journal of Hydraulic Engineering (ASCE), Journal of Hydraulic Research (IAHR), Sedimentology, Acta Geophysica, Journal of Hydro-Environment Research, International Journal of Sediment Research, and Journal of Numerical Mathematics and Stochastics*. He is also an editorial board member of several journals including the *Proceedings of the Royal Society of London A: Mathematical, Physical and Engineering Sciences*. He is a council member of IAHR (2015–2019), member of IAHR Fluvial Hydraulics Committee (2014–), and a past council member of the World Association for Sedimentation and Erosion Research (WASER), Beijing (2010–2013). He is a fellow of the Indian National Science Academy (FNA), Indian Academy of Sciences (FASc), the National Academy of Sciences India (FNASc) and Indian National Academy of Engineering (FNAE). He has also received the award JC Bose fellowship in 2018.

water MDPI

Editorial

Water-Worked Bedload: Hydrodynamics and Mass Transport

Paweł M. Rowiński [1,*] and Subhasish Dey [2]

1   Institute of Geophysics, Polish Academy of Sciences, Ks. Janusza 64, 01-452 Warsaw, Poland
2   Department of Civil Engineering, Indian Institute of Technology Kharagpur, West Bengal 721302, India
*   Correspondence: p.rowinski@igf.edu.pl; Tel.: +48-22-182-60-02

Received: 5 July 2019; Accepted: 5 July 2019; Published: 7 July 2019

Turbulent flow over a natural streambed is complex in nature, especially in the near-bed flow zone, because a natural water-worked bed exhibits a spatially complex, three-dimensional structure [1–3]. This echoes the organization of the bed particles by transport processes. The orientation, imbrication, sorting, and layering of the deposited bed particles are governed by the continual deposition and reworking by several flood cycles. Besides, the bedload transport rate is often predicted from the flow induced bed shear stress with respect to the threshold shear stress, which represents the bed shear stress required for particle entrainment by the flow [4]. Our knowledge on how sediment is transported under such a complex situation is still insufficient, which triggers a good deal of experimental, theoretical, and computational efforts. The near-bed flow is greatly affected by a complex, colossal, fluid–sediment interface giving rise to a spatial flow heterogeneity together with a significant temporal intermittency in the vicinity of the bed. In a natural stream, such a complex flow plays a decisive role in developing its morphological environment. In this process, a so-called *water-worked bed* is formed in a natural stream. By contrast, in laboratory scale experimental studies, a simulated streambed in a flume is generally created by arbitrarily dumping the sediment particle mixture to a given thickness. The sediment bed surface is then worn and levelled, preparing a screeded bed. Even though if the distribution of sediment particle size used in the laboratory experimental study is same as that of the particle size in a natural streambed, the simulated streambed (bed surface characteristics) cannot be deemed to be acceptable as analogous to that in the natural streambed. To be specific, the screeded bed is essentially a mixture of randomly sorted sediment particles and its statistical distribution in terms of bed surface topography is incompetent to mimic a water-worked bed. The bed surface topography and the flow characteristics in water-worked and screeded gravel beds were explored by several researchers [1,5–12]. However, a series of recent studies by Padhi et al. [13–15] indicated a clear distinction in turbulence characteristics in water-worked and screeded gravel bed flows. Therefore, the research on water-worked beds, in addition to the related hydrodynamics and transport processes, being the topic of this Special Issue, demands further attention.

The application of the water-worked bed concept to fluvial hydraulics is developing rapidly and it has already been successful in a number of laboratory scale model studies, including data analysis and interpretation, and supervising conceptual framework and parameterizations by a number of research groups around the world. To be specific, the impact of water-work on transporting sediment, especially as a bedload, is of primary importance. Moreover, sediment transport by the modification of flow at a river protection structure, such as a groyne or a spur dike, has a detrimental effect of forming a scour hole around it. Therefore, the topic of scours at a river protection structure has been a continued interest of research over several decades. Furthermore, investigators have not been restricted to the laboratory scale model studies. They have been, on several occasions, more interested in conducting field studies in natural streambeds, where the beds are water-worked. This tendency in the current research trend is reflected in this Special Issue. It includes nine papers, which can be classified into four categories. The first category is comprised of two review papers from Mrokowska and Rowiński [16] on bedload transport by unsteady flow and Padhi et al. [17] on water-worked gravel beds. These papers deliver an excellent background that is useful for understanding and modeling bedload transport under unsteady flow conditions and for the morphological and flow characteristics in water-worked beds. Both studies are mostly based on experimental investigations. The second category includes studies on river protection structures by Möws and Koll [18] on groynes (one paper) and by Zhang et al. [19,20] on spur dikes (two papers). The former focuses on backwater effect and resistance to flow, and the latter two on scours at spur dikes. Both studies are important from the perspective of designing river protection structures. One paper, by Antico et al. [21], is dedicated to the velocity law in hydraulically rough flow over mobile granular beds, which falls into the third category. The fourth category presents important field studies by Radecki-Pawlik et al. [22] on the Mlynne and Lososina streams in the Polish Carpathians; Huang et al. [23] on the Three Gorges Reservoir (TGR) in China; and Przyborowski et al. [24] on the Jeziorka River and Swider River in Poland.

The Editors finally hope that this Special Issue will be beneficial to advance future research studies and to further develop the water-worked bed concept, including other related issues in laboratory scale models and field studies, and its applications in sedimentology, geophysics, fluvial hydraulics, and environmental and hydraulic engineering.

References

1. Hardy, R.J.; Best, J.L.; Lane, S.N.; Carbonneau, P.E. Coherent flow structures in a depth-limited flow over a gravel surface: The role of near-bed turbulence and influence of Reynolds number. *J. Geophys. Res.* **2009**, *114*, F01003. [CrossRef]
2. McLean, S.R.; Nelson, J.M.; Wolfe, S.R. Turbulence structure over two-dimensional bed forms: Implications for sediment transport. *J. Geophys. Res.* **1994**, *99*, 12729–12747. [CrossRef]
3. Maddahi, M.R.; Afzalimehr, H.; Rowiński, P.M. Flow characteristics over a gravel bedform: Kaj River case study. *Acta Geophys.* **2016**, *64*, 1779–1796. [CrossRef]
4. Dey, S. *Fluvial Hydrodynamics: Hydrodynamic and Sediment Transport Phenomena*; Springer-Verlag: Berlin, Germany, 2014.
5. Kirchner, J.W.; Dietrich, W.E.; Iseya, F.; Ikeda, H. The variability of critical shear stress friction angle, and grain protrusion in water-worked sediments. *Sedimentology* **1990**, *37*, 647–672. [CrossRef]
6. Nikora, V.; Goring, D.; Biggs, B.J.F. On gravel-bed roughness characterization. *Water Resour. Res.* **1998**, *34*, 517–527. [CrossRef]
7. Marion, A.; Tait, S.J.; McEwan, I.K. Analysis of small-scale gravel bed topography during armoring. *Water Resour. Res.* **2003**, *39*, 1334. [CrossRef]
8. Aberle, J.; Nikora, V. Statistical properties of armored gravel bed surfaces. *Water Resour. Res.* **2006**, *42*, W11414. [CrossRef]
9. Buffin-Bélanger, T.; Rice, S.; Reid, I.; Lancaster, J. Spatial heterogeneity of near-bed hydraulics above a patch of river gravel. *Water Resour. Res.* **2006**, *42*, W04413. [CrossRef]
10. Cooper, J.R.; Tait, S.J. The spatial organisation of time-averaged streamwise velocity and its correlation with the surface topography of water-worked gravel beds. *Acta Geophys.* **2008**, *56*, 614–642. [CrossRef]
11. Cooper, J.R.; Tait, S.J. Water-worked gravel beds in laboratory flumes: A natural analogue? *Earth Surf. Proc. Land.* **2009**, *34*, 384–397. [CrossRef]
12. Cooper, J.R.; Tait, S.J. Spatially representative velocity measurement over water-worked gravel beds. *Water Resour. Res.* **2010**, *46*, W11559. [CrossRef]
13. Padhi, E.; Penna, N.; Dey, S.; Gaudio, R. Hydrodynamics of water-worked and screeded gravel beds: A comparative study. *Phys. Fluids* **2018**, *30*, 085105. [CrossRef]
14. Padhi, E.; Penna, N.; Dey, S.; Gaudio, R. Spatially-averaged dissipation rate in flows over water-worked and screeded gravel beds. *Phys. Fluids* **2018**, *30*, 125106. [CrossRef]
15. Padhi, E.; Penna, N.; Dey, S.; Gaudio, R. Near-bed turbulence structures in water-worked and screeded gravel-bed flows. *Phys. Fluids* **2019**, *31*, 045107. [CrossRef]
16. Mrokowska, M.M.; Rowiński, P.M. Impact of unsteady flow events on bedload transport: A review of laboratory experiments. *Water* **2019**, *11*, 907. [CrossRef]

17. Padhi, E.; Dey, S.; Desai, V.; Penna, N.; Gaudio, R. Water-worked gravel bed: State-of-the-art review. *Water* **2019**, *11*, 694. [CrossRef]

18. Möws, R.; Koll, K. Roughness effect of submerged groyne fields with varying length, groyne distance, and groyne types. *Water* **2019**, *11*, 1253. [CrossRef]

19. Zhang, L.; Wang, P.; Yang, W.; Zuo, W.; Gu, X.; Yang, X. Geometric characteristics of spur dike scour under clear-water scour conditions. *Water* **2018**, *10*, 680. [CrossRef]

20. Zhang, L.; Wang, H.; Zhang, X.; Wang, B.; Chen, J. The 3-D morphology evolution of spur dike scour under clear-water scour conditions. *Water* **2018**, *10*, 1583. [CrossRef]

21. Antico, F.; Ricardo, A.M.; Ferreira, R.M.L. The logarithmic law of the wall in flows over mobile lattice-arranged granular beds. *Water* **2019**, *11*, 1166. [CrossRef]

22. Radecki-Pawlik, A.; Kuboń, P.; Radecki-Pawlik, B.; Plesiński, K. Bed-load transport in two different-sized mountain catchments: Mlynne and Lososina Streams, Polish Carpathians. *Water* **2019**, *11*, 272. [CrossRef]

23. Huang, L.; Fang, H.; Ni, K.; Yang, W.; Zhao, W.; He, G.; Han, Y.; Li, X. Distribution and potential risk of heavy metals in sediments of the Three Gorges Reservoir: The relationship to environmental variables. *Water* **2018**, *10*, 1840. [CrossRef]

24. Przyborowski, Ł.; Łoboda, A.M.; Bialik, R.J. Experimental investigations of interactions between sand wave movements, flow structure, and individual aquatic plants in natural rivers: A case study of Potamogeton Pectinatus L. *Water* **2018**, *10*, 1166. [CrossRef]

Review

Impact of Unsteady Flow Events on Bedload Transport: A Review of Laboratory Experiments

Magdalena M. Mrokowska and Paweł M. Rowiński

Institute of Geophysics, Polish Academy of Sciences, Ks. Janusza 64, 01-452 Warsaw, Poland;
m.mrokowska@igf.edu.pl (M.M.M.); p.rowinski@igf.edu.pl (P.M.R.)

Received: 25 February 2019; Accepted: 26 April 2019; Published: 29 April 2019

Abstract: Recent advances in understanding bedload transport under unsteady flow conditions are presented, with a particular emphasis on laboratory experiments. The contribution of laboratory studies to the explanation of key processes of sediment transport observed in alluvial rivers, ephemeral streams, and river reaches below a dam is demonstrated, primarily focusing on bedload transport in gravel-bed streams. The state of current knowledge on the impact of flow properties (unsteady flow hydrograph shape and duration, flood cycles) and sediment attributes (bed structure, sediment availability, bed composition) on bedload are discussed, along with unsteady flow dynamics of the water-sediment system. Experiments published in recent years are summarized, the main findings are presented, and future directions of research are suggested.

Keywords: experiments; flood; hysteresis; river; sediment; bedload; bed shear stress

1. Introduction

Unsteady flow events are intensive phenomena occurring in streams and rivers in various climatic and geomorphic settings [1]. They can be triggered by snowmelts [2], glacial processes, excessive rainfall, dam water releases, or hydropower operations [3,4] and very often entail catastrophic consequences, falling into the category of flood events. Unsteady flows differ in terms of frequency, magnitude, and hydrograph shape and duration, depending on the region and flood origin. Pulsed hydrographs lasting from a few hours to a few days with a steep rising arm [1,5–7] are characteristic for abrupt flows, e.g., dam water releases or flashfloods, while flat hydrographs lasting up to several hundred hours [8,9] are characteristic for flood waves triggered by snowmelt or precipitation.

The quantification of the mobile riverbed response to these changing flow conditions poses a challenge since the effect of temporal flow variability overlaps with the effect of bed structure, bed material composition, and sediment supply. This complexity makes it difficult to separate the effects of flow and the effects of sediment characteristics and availability on bedload transport. Attempts have been made to overcome this difficulty by applying the existing theory of sediment transport in steady flow conditions to unsteady flow problems, e.g., by approximating unsteady flow as a step-wise steady flow. However, this approach has proved to be inadequate in transient flows (dam-break flows, flashfloods) [10]. It is nowadays acknowledged that findings for steady flow cannot be fully transferred to unsteady flow events [11–13] and, as such, a branch of research on sediment transport in unsteady flow conditions has been developing rapidly.

Although unsteady flow events have an enormous impact on fluvial morphodynamics, the academic discussion has still only had a small impact on engineering and water management. The reason for this is because the vast complexity of the problem limits the development of bedload calculation equations that could be applied, e.g., in numerical models. These issues make the topic of sediment transport in unsteady flow conditions one of the most significant and urgent research problems in environmental and engineering hydraulics.

Our knowledge of riverbed morphodynamics and the fate of pollutants [14] during flood events remains insufficient. One reason for this is that the violent character of unsteady flow is a serious constraint preventing field measurements of sediment transport [15]. Nonetheless, some monitoring of bedload in rivers has been conducted and has provided valuable field data [3,7,16]. However, both flow and transport processes are highly variable in time and space, and observations and measurements of detailed processes, such as dynamics of bed morphology during unsteady flow events, still pose a technical challenge.

While safety considerations very often constrain observations in the field, the laboratory assures safe conditions for researchers and apparatus and enables control over measured variables, and, as such, is advantageous over field measurements. Laboratory conditions provide the opportunity to observe and measure detailed processes from reach- to grain-scale, with the capabilities of the measurement equipment being the only limiting factor. There is a certain exception to this in large scale flood experiments, showing, for example, that controlled floods in debris fan-affected canyons of the Colorado River basin redistribute fine sediment and change the local channel morphology by bar-building and bed scour [4]. However, such experiments, although very informative, provide data at a completely different level of accuracy than those discussed in this review. Oscillatory flow experiments simulating sediment transport under waves and currents in coastal zones are another large group of laboratory investigations [17,18], which study grain motion and bedload transport in unsteady flow. However, details of these studies are beyond the scope of this review.

Numerical methods provide another rapidly developing research approach, one which is tightly connected with laboratory data. These numerical methods very often involve a one-dimensional description of the phenomenon due to its smaller numerical cost (see, e.g., Fang et al. [19]), but intensive research has also been conducted on sophisticated 2D numerical methods [20,21]. However, despite the existence of such advanced numerical methods, their progress is limited due to gaps in theory and difficulties in obtaining reliable measurements for calibration. Laboratory studies, in addition to addressing fundamental knowledge gaps, provide the data necessary for the development of numerical models.

Experiments are, therefore, a promising research approach that advances our understanding of sediment transport mechanisms and also complements field and numerical studies. Experimental investigations are indeed in the mainstream of research on sediment transport in unsteady flow since advances in instrumentation and measurement techniques are making it possible to conduct more and more sophisticated experiments [15] that may address challenging research problems. Laboratory studies rarely model conditions in a particular river (a prototype). Instead, they usually have a general context and aim to identify the mechanisms underlying fundamental processes [11].

The literature on laboratory experiments touches upon a number of detailed problems, to be addressed further on in this paper, from grain-scale to bulk transport processes, additionally complicated by the temporal and spatial variability of water flow. This may give the impression that the state of research in the field is chaotic; hence, we believe that overviews of specific areas of this complex topic will be useful. Laboratory research on sediment transport in unsteady flow has been summarized to some extent in a few review papers. They focused on sediment transport characteristics in relation to pollutant transport in unsteady flow [10], factors affecting the hysteretic relationship between flow rate and sediment transport [22], and presented current laboratory techniques applied in bedload studies, both in steady and unsteady flow conditions, and dedicated a few sections to the impact of sediment supply, armoring, and hydrograph on bedload transport in unsteady flow [15].

The present review focuses on the transport of coarse-grain bedload from the perspective of experimental studies, and the aim is to summarize existing directions in laboratory research on sediment transport in unsteady flow conditions and to point out future perspectives for experimental investigations developing this research topic. This review does not aim to be exhaustive and focuses on selected issues: (1) to summarize recent laboratory studies in terms of experimental conditions and modeling issues; (2) to present existing interpretations of the hydrodynamics of unsteady flow; (3) to present current knowledge on the interaction between unsteady flow and riverbed, and (4) to discuss the research questions addressed in previous studies and future needs and perspectives. Laboratory studies are presented within the wider context of sediment transport research including field, numerical, and theoretical studies since experiments are inherently connected with these researches. They all contribute to the understanding of bedload transport processes and generate new research questions that may become topics of laboratory experimentation.

2. Experimental Conditions

Laboratory experiments are designed so that certain, independent variables can be controlled to assess their impact on other, dependent variables, which is not possible in field observations. Hydrograph characteristics, initial bed composition and structure, and sediment supply are usually taken as independent variables, and their effect on sediment transport characteristics is measured. The number of combinations is endless, and Table 1 summarizes some of the experimental conditions considered in recent studies, with a focus on experiments looking at coarse-grain and bi-modal bed material composition.

The hydrographs applied in experimental studies vary in shape, duration, flow magnitude, time-to-peak flow, and proportion between rising and falling limb duration, so as to model various types of flood waves occurring in natural conditions (Figure 1). Triangular [13], trapezoidal [23], and step-wise [24–26] hydrographs have been considered. Other researchers have designed more naturally-shaped hydrographs in the form of smooth curves [27–29]. The duration of hydrographs has varied from a few minutes [12] to a few hours [5]. Cycled hydrographs have been designed to model the influence of successive floods or other unsteady flow events on the bed texture and sediment transport [11,25,27,30–33].

The bed of experimental channels has been composed of unimodal sand or gravel [11,13,27,34], sand–gravel [12,25,35], silt-gravel and silt-sand mixtures [26,32], and tri-modal sand–gravel mixtures [36]. An idealized bed structure, i.e., well-mixed and screeded, has been applied to exclude the influence of initial bed morphology prior to single [24,34] and cycled hydrographs [32]. A bed water-worked by antecedent flow prior to a single hydrograph has been applied to simulate conditions similar to those in nature [12,23,28]. Other studies have combined structured water-worked gravel beds with cycled hydrographs [31]. Some studies have applied a more complex planar morphology, for instance, to simulate alternate bar topography [27].

Sediment supply has been controlled in laboratory studies to simulate sediment feeding or sediment starving conditions [24,30]. Sediment-feeding conditions occur when a sediment load from upstream is provided, and the rate of supply is larger than the bedload transport capacity; in the converse, as in the case of flow below dams, sediment-starved conditions occur. Sediment supply also has a technical motivation, as a way to control scour and deposition during an experiment when the variation of bed level is undesirable [11,30,31,34]. Erosion processes may considerably affect the water surface level when the water depth is relatively small in laboratory flumes [34].

Table 1. Laboratory studies on sediment transport under unsteady flow conditions.

Study	Type of Hydro-Graph [1]	Channel Dimensions [2] and Slope	Flow	Initial Bed Conditions	Sediment and Supply	Hyste-Resis [3]
Bombar et al., 2011 [23]	S, triangular, trapezoidal	18.6 × 0.8 × 0.75 slope: 0.005	peak about 80 L/s duration: 67–270 s	screeded and water-worked	gravel, d_{50} = 4.8 mm	N/A
Curran et al., 2015 [37]	S, stepped	11 × 0.6 × 0.5	duration: 76 min	well-mixed, screeded	70% sand, 30% gravel; d_{50} = 0.5 mm; sediment recirculation	N/A
Ferrer-Boix, and Hassan. 2015 [31]	S, pulsed	18 × 1 × 1 slope: 0.022	variable duration (1–10 h) low flow 0.065 m²/s, followed by 1.5 h constant high flow pulse 0.091 m²/s	water-worked	d_{mean} = 5.65 mm; 20% sand; constant feed rate 2.1 g/m/s	N/A
Guney et al., 2013 [12]	S, triangular	18.6 × 0.8 slope: 0.006	base flow: 9.5 L/s; peak flow: 49.6 L/s; duration: 10 min	well-mixed, water-worked	gravel/sand mixture; d_{50} = 3.4 mm, no supply	C, CC
Hassan et al., 2006 [5]	S, stepped triangular	9 × 0.6 × 0.5	0.012–0.055 m³/s; duration: 0.83–64 h	water worked	range of grain size: 0.180–45 mm; no supply	N/A
Humphires et al., 2012 [27]	S, naturally-shaped (lognormal)	28 × 0.86 × 0.86	peak flow: 35 L/s, 25 L/s; duration: 14.5 h, 8.5 h	armored	d_{50} = 4.1 mm sediment pulses	S
Lee et al., 2004 [13]	S, triangular	21 × 0.6 × 0.6 slope: 0.002	base flow: 0.04 m²/s; peak flow 0.05–0.14 m²/s; duration: 21–80 min		d_{50} = 2.08 mm no supply	CC
Li et al., 2018 [29]	S, naturally-shaped (smooth sinusoidal curves)	35 × 1.2 × 0.8 slope: 0.003	peak flow 0.018 m²/s and 0.038 m²/s		gravel (2–4 mm), sand (0.1–2 mm), 100% gravel; 100% sand; 53% gravel and 47% sand; 22% gravel and 78% sand; constant feed rate 2.1 g/(m s)	N/A
Mao, 2012 [24]	S, stepped symmetrical	8 × 0.3 slope: 0.01	0.024–0.085 m²/s	mixed and screeded sediment	20% sand, 80% gravel, d_{50} = 6.2 mm, continuous recirculation	C
Mao, 2018 [25]	C, three types of stepped symmetrical	8 × 0.3 slope: 0.01	0.024–0.085 m²/s	water-worked by steady antecedent flow	20% sand, 80% gravel, d_{50} = 6.2 mm, supply	C, CC
Martin and Jerolmack, 2013 [38]	S, pulsed and triangular	15 × 0.92 × 0.65 slope: 0	peak flow: 81.4, 111.7 L/s; low flow: 39.1, 63.3 L/s; duration: several hours	water-worked by low flow	d_{50} = 0.37 mm no supply	N/A
Mrokowska et al., 2018 [34] Mrokowska et al., 2016 [39]	S, triangular	12 × 0.49 × 0.6 slope: 0.0083	base flow: 0.0035–0.0131 m³/s; peak flow: 0.0387–0.0456 m³/s; duration: 400–800 s	well-mixed, screeded, without and with antecedent flow	d_{mean} = 4.93 mm supply	C
Nelson et al., 2011 [40]	S, square-wave	6 × 0.25 × 0.4 slope: 0.002	peak: 0.02 m³/s	well-sorted	sand d_{50} = 0.58 mm no supply	N/A
Orru et al., 2016 [36]	S, one step	14 × 0.4 × 0.45 slope: 0.0022	stepped increase form 0.0465 m³/s to 0.0547 m³/s	water-worked	tri-modal sediment mixture d_{50} = 1 mm, d_{50} = 6 mm, d_{50} = 10 mm; no supply	no
Perret et al., 2018 [26]	C, stepped symmetrical	18 × 1 × 0.8 slope: 0.01	-	loose and packed gravel beds, infiltrated with fine grains	gravel d_{50} = 6.8 mm and bimodal gravel–sand and gravel–silt	N/A

Table 1. *Cont.*

Study	Type of Hydro-Graph [1]	Channel Dimensions [2] and Slope	Flow	Initial Bed Conditions	Sediment and Supply	Hyste-Resis [3]
Phillips et al., 2018 [11]	C, four different shapes: triangular and rectangular	30 × 0.5	-	-	unimodal well-mixed, d_{mean} = 7.2 mm	N/A
Piedra et al., 2012 [41]	S, stepped, increasing discharge	7 × 0.9 slope: 1/150	peak: 29–34 L/s	-	gravel d_{50} = 6.6 mm no supply	No
Redolfi et al., 2018 [30]	C, square-wave and triangular	24 × 2.9, 24 × 0.8 slope: 1.0%	square-wave: 1.2–2.5 L/s, 1.5–2.5 L/s;triangular: 0.5–2.5 L/s	well-sorted sand, water-worked by antecedent low flow	sand d_{50} = 1 mm supply	C
Shvidchenko and Kopaliani, 1998 [42]	S, stepped	outdoor plot: 84 × 10; flume: 100 × 1; recirculating tilting flume: 18 × 2.46	-	braided channel	d_{mean} = 0.69 mm d_{max} = 5–8 mm recirculating flume: d_{50} = 4.3 mm	No
Waters and Curran, 2015 [32]	C, stepped	9 × 0.6 × 0.5	duration: 76 min, cycled with 2 h base flow between, peak flow: 0.073, 0.131 m²/s, base flow 0.029 m²/s	well mixed screeded flat, antecedent low flow	70% sand, 30% gravel, d_{50} = 0.55 mm and 70% sand, 30% silt, clay d_{50} = 0.27 mm no supply	F8, CC most frequent
Wang et al., 2015 [28]	S, natural-shaped	8 × 0.3 × 0.3 slope: 0.0083	base flow 8 L/s, peak flow 13.5–18 L/s; duration: 120–141 s	screeded, antecedent flow	range of grain size: 1–16 mm; d_{50} = 5 mm, unimodal and bimodal	C
Wong and Parker, 2006 [33]	C, triangular	22.5 × 0.5	peak flow: 0.065–0.102 m³/s; duration 15–60 min	well-sorted	gravel, d_{50} = 7.1 mm, constant feed	N/A

[1] S—single, C—cycled; [2] length (m) × width (m) × depth (m); [3] Hysteresis in the relationship between total sediment transport rate and flow rate; C—clockwise, CC—counterclockwise, F8—figure-8 shape.

Undistorted mobile bed models based on Froude similitude have usually been applied to model sediment transport in unsteady flow. A general rule applies to unsteady flow experiments: When fully rough flow occurs in a river; it is enough to assure that scaled flow is also fully rough to satisfy the Reynolds number criterion. Then the Froude number becomes the main criterion to calculate scaling between the model and the prototype [43]. The similitude of boundary shear stresses is usually obtained by applying the Shields number to satisfy the similarity of forces acting on sediment particles in a prototype and a model [43]. Mao [24] used Froude scaling to prepare a model that represents a narrow gravel-bed river. The model at a scale 1:30 represented a 10-m wide stream with a bed composed of material with d_{50} = 200 mm, while flow corresponded to a flashy flood lasting 10 h and a snowmelt flood lasting 83 h. Redolfi et al. [30] constructed a model representing a typical gravel bed river with d_{50} = 50 mm and flood duration of 1 h in a model corresponding to 7 h in a prototype. Shvidchenko and Kopaliani [42] provided in-depth theoretical commentary on similitude laws and their study presented a model of Laba River at a scale of 1:50. Lee et al. [13] commented on the applicability of Froude similitude to hydrograph design, concluding that this law can be adopted in unsteady flow even if equilibrium conditions of bed morphology are not met.

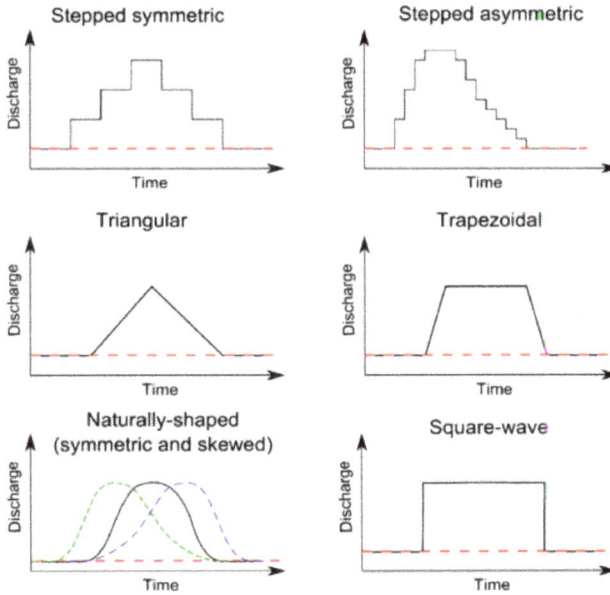

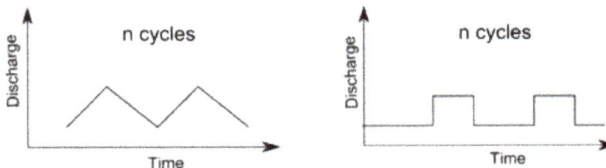

Figure 1. Main types of unsteady flow hydrographs tested in laboratory experiments. The dotted red line denotes a base flow.

3. Hydrodynamic Aspects of Sediment Transport

Bed material is set into motion as the result of forces exerted by flowing water, usually represented by drag and lift on a grain scale and friction on a larger scale. Thus, proper evaluation of these forces is necessary to assess sediment transport. Forces acting on sediment depend on the flow attributes, such as turbulence characteristics and mean flow characteristics. A variety of studies in steady flow conditions have shown that there is mutual interaction between flow properties and movement of sediment, demonstrating that flow properties are modified in the presence of bedload as compared to the clear water (an absence of sediment transport) counterpart [44–46]. For example, near-bed velocity fluctuations, and consequently Reynolds shear stresses, diminish when the channel bed is movable [47].

A better understanding of grain scale mechanics is necessary to improve the assessment methods of bed load transport [48]. Detailed laboratory measurements have demonstrated the impact of pressure gradients around grains and turbulence events on the entrainment of sediment grains in steady flow [15]. Although unsteadiness is an immanent feature of river systems, its impact on the fate of single particles has yet to be sufficiently understood. But one has to acknowledge the attempts to understand the influence of unsteadiness (hydrograph characteristics) on particular forces, for example, the magnitude of lift [49]. Another related example is a study considering the effect of turbulent flow parameters on the movement of particles in natural conditions, which has shown that flow acceleration affects bedload transport [50].

Contrasting results have been reported concerning the impact of sediment transport on flow resistance. On the one hand, bedload transport has been found to enhance flow resistance, due to additional flow energy dissipation through interactions between sediment grains and the extraction of momentum from the flow [51]. On the other, however, a large body of research has reported the reverse trend, showing decreased flow resistance or a negligible effect in movable bed conditions [44,52,53]. Since flow resistance varies due to the evolution of bed structure as water flows over movable bed, it has been proposed to apply a flow-dependent roughness factor instead of a fixed roughness coefficient to calculate bedload using resistance equations [54]. Our understanding of the abovementioned phenomena in steady flow is still incomplete and much less is known about flow resistance in unsteady flow with a movable boundary.

It has been well known that flood hydrograph characteristics affect sediment transport capacity through time-variable bed shear stresses. Recently, some progress has been made with methods to evaluate bed shear stresses in unsteady flow conditions [55–58], but these are mostly indirect methods. It is quite unfortunate that the techniques used to measure instantaneous values of shear stresses are not well developed [59]. Flow resistance in unsteady flow has been widely considered using bed shear stress τ (N/m^2) or friction (shear) velocity u_* (m/s) (Equation (1)) to quantify friction.

$$u_* = \sqrt{\frac{\tau}{\rho}} \tag{1}$$

where ρ—water density (kg/m^3).

The first report on the friction velocity in unsteady flow over a rough gravel bed was presented by Tu and Graf [60]. They found that friction velocity achieves the peak value along a rising limb before peaks of water depth and discharge, and that bed shear stress is larger along the rising limb than along the falling limb of the hydrograph (Figure 2). Similar results were later obtained by Graf and Song [56,61] and Nezu and Nakagawa [62]. Although this relationship was observed for immobile bed conditions, it appears to be significant for research on bedload transport. Since the peak of bedload is most likely around the peak of bed shear stress, the most intensive sediment transport is expected before the discharge peak, provided sediment is available. It is also more likely that armored bed is destroyed in the region of increased shear velocity.

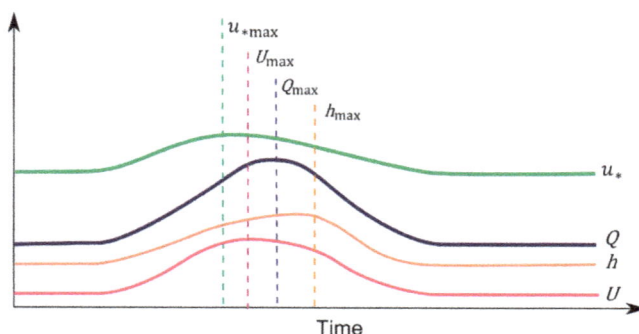

Figure 2. Schematic representation of temporal variation of flow parameters, indicating the sequence of peak values. Based on [56,61], u_{*max}—maximum friction velocity (m/s), U_{max}—maximum mean flow velocity (m/s), Q_{max}—maximum flow rate (m/s), h_{max}—maximum flow depth (m).

The shear velocity in unsteady flow conditions may be conveniently evaluated from the following formula:

$$(u_*)_{SV} = \left[gR\left(I + \frac{U}{gh}\eta + \left(\frac{U^2}{gh} - 1 \right)\vartheta - \frac{1}{g}\zeta \right) \right]^{\frac{1}{2}} \qquad (2)$$

where $\eta = \frac{\partial h}{\partial t}, \vartheta = \frac{\partial h}{\partial x}, \zeta = \frac{\partial U}{\partial t}$, t—time (s), U—mean cross-sectional velocity (m/s), x—horizontal spatial coordinate (m). This formula may be derived from flow equations—the momentum conservation equation and the continuity equation in the form of the Reynolds 2D model [56] or 1D Saint–Venant model [60]. It should be kept in mind that the Saint-Venant model was derived for immobile bed conditions, and the movement of bed elements may introduce some degree of uncertainty. Generally, there is a high level of ambiguity in the definition of bed shear stresses in a mobile channel boundary [63,64].

Equation (2) may be simplified in a number of cases by neglecting particular terms. For rapidly varied flows, which may occur in the case of dam-break flows or ephemeral floods, all terms of the equation are significant. However, in many cases of seasonal floods and related laboratory models, the acceleration terms are significantly smaller than the others, and they may be removed; for further information see, e.g., Mrokowska et al. [57]. Shear velocity has been evaluated with formula derived from flow equations in a number of unsteady flow studies [57,58,65–67] as well as in studies on sediment transport in unsteady flow conditions [12,28,34].

It should be noted that water surface slope $(I - \vartheta)$ is present in each form of the equation irrespective of the simplifying assumptions. Water surface slope has a pronounced impact on sediment transport assessment especially in the case of high-yield ephemeral streams [68]. At the same time, this variable is difficult to control in laboratory mobile bed conditions, where water surface fluctuations tend to occur in relatively shallow flow [34].

4. Impact of Unsteady Flow on Bed Structure and Composition

The texture of the riverbed may evolve rapidly during floods due to the combined effect of variable shear stresses and the availability of sediment grains. When coarse-grained riverbed is examined in natural conditions, it is almost impossible to distinguish which aspects of bed structure are the effect of steady or unsteady flow [11], since it has been shown that both types of flow trigger the same fundamental phenomena involving grain organization. An example is the formation of an armor layer (i.e., a bed surface layer of grains coarser than subsurface material). Three well-documented mechanisms of armor formation are horizontal downstream preferential transport of finer grains (winnowing), kinematic sieving, moving grains in a vertical direction, and spontaneous percolation when the coarse fraction is immobile [15,69]. A recent laboratory study considered the armoring process in the context of dense granular flow and found that vertical segregation of grains may occur not only due to the action of flowing water but also due to granular bottom-up segregation [70].

Much attention has been paid to the formation and persistence of an armor layer in steady and unsteady flows. It has been shown that low steady flow promotes channel bed consolidation and the formation of an armor layer [15,71]. However, experimental research has demonstrated that not only steady but also unsteady flows may trigger the formation of armor conditions [5]. An armoring effect has been identified for flat flood waves (as in snowmelt floods) with a long falling limb. Another laboratory study has shown the formation of coarse-grain clusters under increasing discharge [41]. A similar stabilizing effect has been demonstrated for a low magnitude hydrograph occurring before another flood wave [25]. It has been demonstrated that armoring is more likely in sediment starving conditions than when sediment is available [5]. Thus, antecedent steady flow without sediment feeding is usually applied in the laboratory to prepare an armored bed for unsteady flow experiments [12,27,36].

Bertin and Friedrich [72] reported that total mobilization of coarse sediments is only possible during high-magnitude floods [73], indicating that partial transport promoting stable armor layer formation prevails in coarse-grained riverbeds. Armor layer destruction has been observed for peaky dam water release hydrographs with high variability of flow and large stream power [74].

Various effects of unsteady flow on bed composition have been reported in the literature. Some field studies have indicated that bed composition after a flood event remained the same as before the flood [75,76]. A possible explanation is the constant supply of sediment from upstream enabling mobile armor layer formation [75]. Conversely, the laboratory experiments presented in Mao [24] showed that bed surface composition and arrangement were modified after a simulated flood event with a mobile armor layer.

It has been demonstrated that the stability of an armor layer depends on the grain size distribution of the sediment supply, with a mobilizing effect when grains finer than the bed material are provided [15,77]. Mobility of grains changes considerably in bi-modal sediments. Steady flow experiments have demonstrated that the threshold for gravel transport is reduced when sand is added [78,79]; when sand content exceeds 35 to 40% then sediment behaves like sandy material [80]. Similar effects have been revealed in unsteady flow experiments. Li et al. [29] demonstrated that transport of gravel is higher in a gravel–sand mixture than in a pure gravel bed and the transport of sand is lower due to a hindering effect. Wang et al. [28] compared the total sediment transport rate for unimodal sediments and bi-modal gravel–sand mixture and found that the mixture is transported at higher bedload rates than the unimodal counterpart. Moreover, experiments with sand–silt and gravel–sand mixtures have demonstrated that bedload transport is larger in the first case since gravel has a stabilizing effect on the second mixture [32]. Perret et al. [26] investigated the effect of infiltration of sand and silt experimentally into a gravel matrix on sediment transport in an unsteady flow event. Their approach differed from that of previous studies in that they did not use mixed sediment but instead infiltrated fine grains into the gravel matrix. They reported that cohesive sediment consolidates the bed and, thus, reduces sediment flux compared to bed composed of gravel, whereas infiltrated sand enhances sediment transport.

5. Total and Fractional Bedload Transport

The impact of flow unsteadiness on sediment transport manifests itself in the total weight of sediment transported during a flood (total sediment yield). Laboratory experiments comparing total yield for a given hydrograph and for the equivalent-volume steady flow have shown that sediment yield is higher for unsteady flow than for the equivalent steady flow counterpart, indicating that unsteadiness enhances sediment transport [13,28,29]. It has been demonstrated that total sediment yield was up to an order of magnitude higher for the naturally-shaped hydrograph (peak flow rate 18 L/s and duration 7200 s) than for the equivalent steady flow (flow rate 13.45 L/s and duration the same as for the unsteady flow hydrograph) [28]. Li et al. [29] presented data extending these observations and reported, among other findings, that for a naturally-shaped unsteady flow hydrograph (duration 7 h and flow rate peak 0.038 m^2/s) and its volume-equivalent steady flow counterpart, total yield of unimodal sediment decreased from 114.8 kg to 11.4 kg for gravel and from 440.2 kg to 271.6 kg for sand. However, Yager [15] reported a few studies that found smaller total yield during a flood event than in equivalent steady state conditions. This is indicative of the fact that other factors, such as sediment availability, have to be considered in addition to the flow characteristics in such comparisons.

Wang et al. [28] claimed that flow unsteadiness and hydrograph magnitude, and not hydrograph shape, are major factors influencing sediment transport yield, while Redolfi et al. [30] found hydrograph magnitude and shape to be the main factors influencing the averaged bedload transport.

The structure of the bed surface, sediment supply conditions, and unsteadiness of flow seem to be major factors affecting temporal bedload and fractional transport [28,32]. Bedload rate varies in time during flood wave propagation, exhibiting hysteresis in the relationship between total sediment transport rate and flow rate with time lag effects. Both a clockwise hysteresis, with the bedload peak preceding the discharge peak, and a counterclockwise hysteresis, with the reverse trend, have been observed in nature and in the laboratory, and a more complicated figure-8 shape has also been reported (see Table 1 and Figure 3).

Various factors affect the loop-shaped relationship between bedload rate and discharge, e.g., the availability of sediment, including supply from upstream and the organization of bed sediments. In-depth analysis of these issues can be found in a review by Gunsolus and Binns [22]. A clockwise hysteresis has been reported most often in intact coarse-bed conditions [12,24,25,27,28,30,34]. A counterclockwise hysteresis has been observed mainly for initial armored or well-sorted conditions, where the bedload peak occurred after the breakup of consolidated material along the receding limb of the hydrograph [12,32] or, as in the case of sand [13], where it has been associated with bed forms.

Laboratory studies on bi-modal gravel–sand sediments have revealed the variation in grain size transported during a flood event, i.e., fractional transport [12,24,25,28]. The clockwise bedload–discharge relationship reported for these mixtures has been accompanied by a counterclockwise hysteresis in fractional bedload transport, that is, finer grains predominate in sediment transported along the rising limb and coarser grains dominate along the receding one. Conversely, a counterclockwise hysteresis in total bedload appeared along with clockwise loop in fractional transport [12].

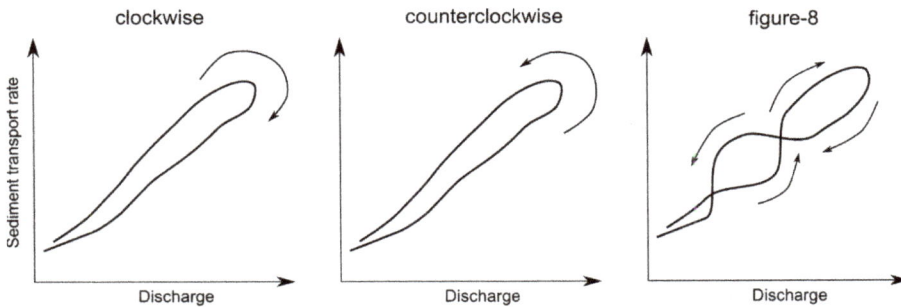

Figure 3. Types of bedload transport hysteresis. Arrows denote hysteresis direction.

Although existing data are not sufficient to quantify the effect of hydrograph shape on hysteresis in the bedload rate–discharge relationship [22], it is important to stress the significance of flow unsteadiness itself since the literature reports that hysteresis has been observed only for rapidly changing flows [11]. Wang et al. [28] demonstrated that unsteadiness affects both total and fractional bedload. Their findings show that hydrographs with a higher rate of unsteadiness (short peaky flows) transport relatively more sediment than gradual events. More significantly, the authors have suggested that in these flashy hydrographs, sediment transport may be initiated for lower flow than in the case of flat hydrographs. These observations are in line with the theory of unsteady flow, which says that bed shear stress achieves its peak along a rising limb (see Section 3), thereby making favorable hydraulic conditions for maximum bedload transport. Whether a bedload peak appears around the peak of bed shear stress or not depends on other sediment and bed-related factors, e.g., sediment availability.

6. The Impact of Flood History on Bedload Transport

Field observations have shown that single unsteady flow events may considerably modify riverbed morphology [81]. However, riverbeds evolve continuously as a result of successive periods of low flow disrupted by flood events, or repeating flashfloods, as for instance occur in ephemeral streams in arid regions [1,6,7], or pulsed flows below dams [3]. The cycles of low and high flows, i.e., flood history, have a significant influence on riverbed morphodynamics on a long-term scale [3,11,25].

Laboratory studies on the impact of preceding flow events on sediment transport during a flood have been increasingly reported in the literature in recent years. In this approach, multiple flood events are considered to check what memory a fluvial system may exhibit. A sequence of floods varying in duration, magnitude, shape, and intraflood flow have been simulated in laboratory channels to study the effects of previous flows on sediment grain arrangement in gravel or bi-modal riverbeds. Hassan et al. [5] claimed that a few cycles of flat hydrographs per year promote armoring. Experimental studies on the effect of cycled pulsed hydrographs have shown downstream fining of surface sediment and demonstrated that bed structure and bedload transport are affected by the frequency of unsteady flow and duration of low flow periods [31]. Similarly, Redolfi et al. [30] observed downstream fining and highlighted the role of low flow periods when finer material is winnowed.

Mao [25] ran a sequence of stepped hydrographs of different magnitudes on a bi-modal bed composed of gravel–sand mixture to show that high magnitude hydrographs affect the sediment transport rate during subsequent high- and low-magnitude flood events, while the preceding low-magnitude hydrograph affects the sediment transport rate only when it is followed by a low-magnitude event. The study indicated potential reasons for the reduction of the sediment transport rate during a subsequent hydrograph when the first one has a high magnitude: (1) mobilization of coarser grains from the thicker layer of active sediments and (2) kinematic sieving reducing the availability of fine grains. The reduction of sediment transport during a low-magnitude event when it is preceded by a low-magnitude event was attributed to (1) the formation of clusters and patches which stabilize the bed and (2) the effect of coarse grain protrusion. The stabilizing effects of antecedent low-flow have been observed earlier in studies on single hydrographs with low antecedent flows, e.g., in Waters and Curran [32]. Mao [25] also reported decreasing bedload hysteresis through successive flood events and associated it with the vertical winnowing of fine grains. Phillips et al. [11] performed an experiment with unimodal sediment and observed no memory effects, which, along with the studies mentioned above, indicates that sediment composition seems to be a major factor contributing to the memory of a fluvial system.

7. Recapitulation, Open Questions, and Outlook

It should be evident from the material discussed above that understanding bed load transport under unsteady flow conditions is central to understanding the impact of flood events in water courses. Bearing in mind the complexity of the physics underlying these processes, especially when temporal changes are taken into account, we argue that the best approach to increase our knowledge of bed load transport involves laboratory flume experiments. While there are opportunities to move forward at increased pace, there are also significant challenges faced by hydraulic researchers, which have been discussed in this paper.

In principle this paper has sought to summarize the current knowledge on the dynamics of bed load transport and interactions between flow unsteadiness and riverbed. This review has discussed only a few selected topics, a selection necessarily biased by the interests and/or involvement of the authors. As mentioned in the introduction, a few reviews of this topic already exist, and they are recommended for a more complete overview of the field.

Significant advancements have been made, particularly in the last ten years, in understanding the impact of flood events in watercourses with gravel or bi-modal sediment. One reason for such interest in the subject is the practical significance of unsteady flow events in mountain regions, which are prone to flooding and serious alteration of fluvial system morphodynamics. Another reason is the eagerness to solve the complex fundamental two-phase flow problem involving water and grains with multimodal size distribution in unsteady flow conditions. Sediment transport in sand bed rivers, where the evolution of bed forms instead of grain organization affects the transport rate, has gained much less attention; however, this trend seems to be changing [82]. Even with the mentioned advancements, challenges remain, and we are far from being able to posit possible generalizations due to the limited number of various conditions studied and too few experiments performed under similar conditions.

We can, however, point out some of the limitations of the works described in this paper. First, we have assumed the sediment to be cohesionless, and we have discussed only research dealing with such situations, although we realize that such a condition does not have to be satisfied in many natural settings, particularly when the bed is covered by clay or mud. A second major assumption concerns the quite artificial shape of most of the hydrographs created in laboratory flumes. However, the variation in those hydrographs, in terms of such factors as shape, duration, and flow magnitude, does provide insight into their impact on bed load transport.

Almost every published study has brought a few open questions showing how much effort is still necessary to gain a basic understanding of sediment transport processes on a local grain scale, a reach scale, and on the scale of the whole basin. Based on those studies and our own experience and intuition, we may point to the following goals and directions of future experimental studies. There are often issues that are conceptually very straightforward but would still pose technical problems; a good example is the experimental evaluation of the gradient of flow depth and consequently friction velocity and the bed shear stress. This is associated with difficulties in avoiding water surface slope fluctuations under unsteady flow conditions in a flume. New techniques allowing the above to be solved are still in great need. If this can be satisfactorily resolved, the next task of crucial importance seems to be quantifying the effect of flow unsteadiness and variable shear stresses during the hydrograph on the sediment transport rate and hysteresis in the relationship between bedload rate and the discharge.

It is expected that trials to elucidate how sediment transport dynamics in an unsteady flow event depend on the flow memory of the system (e.g., associated with a flood event sequence) remains an active area of research for the foreseeable future. After all, we realize that the alterations in flood patterns forecast by climate change models [83,84] may enhance the variability of unsteady flow events and intraflood conditions, and the complexity of flow-sediment interactions may increase. It may be equally important to quantify the effect of kinematic sieving occurring in multi-modal grain size distributed sediment during a sequence of flood events [25].

We still do not know if it is possible to quantify sediment transport in unsteady flow given the sediment composition and characteristics of transient flow conditions—so far we are able to qualitatively observe the relationships between various quantities. Going forward, it is crucial to study how various supply conditions in terms of supply rate and grain size affect sediment transport during a flood event and a sequence of events [25]. Apart from its basic importance, this issue has practical implications for experiment design: The effect of initial bed conditions is evident in the analyses of the total supplied and transported sediment mass. Most experiments do not include complex bed configuration involving bed forms and alternate bars, but such extension seems to be absolutely necessary if we want to apply our knowledge to real rivers. Along these lines, Perret et al. [26] pointed to the study of the effect of multimodal sediment composition, considering the effect of fine sediment infiltration into the gravel matrix.

We are also convinced that the major problem in our understanding of bed load transport under unsteady flow conditions lies in the limited understanding of the hydrodynamics of unsteady flows and its interrelation with mass transport phenomena. Therefore, much attention has to be paid to quantifying the effect of flow unsteadiness and variable shear stresses during a given hydrograph on the sediment transport rate and hysteresis in the relationship between bedload rate and discharge, to studying the mutual influence between turbulence and sediment transport in unsteady flow and, last but not least, to correlating flow velocity distributions with bed load discharge.

Author Contributions: Conceptualization, M.M.M and P.M.R.; methodology, M.M.M.; resources, M.M.M and P.M.R.; writing—original draft preparation, M.M.M.; writing—review and editing, M.M.M. and P.M.R.; visualization, M.M.M.; supervision, P.M.R.

Funding: This work was supported within statutory activities No 3841/E-41/S/2019 of the Ministry of Science and Higher Education of Poland.

Conflicts of Interest: The authors declare no conflict of interest.

References

1. Fielding, C.R.; Alexander, J.; Allen, J.P. The role of discharge variability in the formation and preservation of alluvial sediment bodies. *Sediment. Geol.* **2018**, *365*, 1–20. [CrossRef]
2. Millares, A.; Polo, M.J.; Monino, A.; Herrero, J.; Losada, M.A. Bed load dynamics and associated snowmelt influence in mountainous and semiarid alluvial rivers. *Geomorphology* **2014**, *206*, 330–342. [CrossRef]
3. Aigner, J.; Kreisler, A.; Rindler, R.; Hauer, C.; Habersack, H. Bedload pulses in a hydropower affected alpine gravel bed river. *Geomorphology* **2017**, *291*, 116–127. [CrossRef]
4. Mueller, E.R.; Schmidt, J.C.; Topping, D.J.; Shafroth, P.B.; Rodriguez-Burgueno, J.E.; Ramirez-Hernandez, J.; Grams, P.E. Geomorphic change and sediment transport during a small artificial flood in a transformed post-dam delta: The Colorado River delta, United States and Mexico. *Ecol. Eng.* **2017**, *106*, 757–775. [CrossRef]
5. Hassan, M.A.; Egozi, R.; Parker, G. Experiments on the effect of hydrograph characteristics on vertical grain sorting in gravel bed rivers. *Water Resour. Res.* **2006**, *42*. [CrossRef]
6. Billi, P. Flash flood sediment transport in a steep sand-bed ephemeral stream. *Int. J. Sediment Res.* **2011**, *26*, 193–209. [CrossRef]
7. Reid, I.; Laronne, J.B.; Powell, D.M. Flash-flood and bedload dynamics of desert gravel-bed streams. *Hydrol. Process.* **1998**, *12*, 543–557. [CrossRef]
8. Sui, J.; Koehler, G.; Krol, F. Characteristics of Rainfall, Snowmelt and Runoff in the Headwater Region of the Main River Watershed in Germany. *Water Resour. Manag.* **2010**, *24*, 2167–2186. [CrossRef]
9. Kampf, S.K.; Lefsky, M.A. Transition of dominant peak flow source from snowmelt to rainfall along the Colorado Front Range: Historical patterns, trends, and lessons from the 2013 Colorado Front Range floods. *Water Resour. Res.* **2016**, *52*, 407–422. [CrossRef]
10. Tabarestani, M.K.; Zarrati, A.R. Sediment transport during flood event: A review. *Int. J. Environ. Sci. Technol.* **2015**, *12*, 775–788. [CrossRef]
11. Phillips, C.B.; Hill, K.M.; Paola, C.; Singer, M.B.; Jerolmack, D.J. Effect of Flood Hydrograph Duration, Magnitude, and Shape on Bed Load Transport Dynamics. *Geophys. Res. Lett.* **2018**, *45*, 8264–8271. [CrossRef]
12. Guney, M.S.; Bombar, G.; Aksoy, A.O. Experimental Study of the Coarse Surface Development Effect on the Bimodal Bed-Load Transport under Unsteady Flow Conditions. *J. Hydraul. Eng.* **2013**, *139*, 12–21. [CrossRef]
13. Lee, K.T.; Liu, Y.L.; Cheng, K.H. Experimental investigation of bedload transport processes under unsteady flow conditions. *Hydrol. Process.* **2004**, *18*, 2439–2454. [CrossRef]
14. Muirhead, R.W.; Davies-Colley, R.J.; Donnison, A.M.; Nagels, J.W. Faecal bacteria yields in artificial flood events: Quantifying in-stream stores. *Water Res.* **2004**, *38*, 1215–1224. [CrossRef]
15. Yager, E.M.; Kenworthy, M.; Monsalve, A. Taking the river inside: Fundamental advances from laboratory experiments in measuring and understanding bedload transport processes. *Geomorphology* **2015**, *244*, 21–32. [CrossRef]
16. Rickenmann, D. Variability of Bed Load Transport during Six Summers of Continuous Measurements in Two Austrian Mountain Streams (Fischbach and Ruetz). *Water Resour. Res.* **2018**, *54*, 107–131. [CrossRef]
17. Hallermeier, R.J. Oscillatory bedload transport: Data review and simple formulation. *Cont. Shelf Res.* **1982**, *1*, 159–190. [CrossRef]
18. Ribberink, J.S.; Katopodi, I.; Ramadan, K.A.H.; Koelewijn, R.; Longo, S. Sediment transport under (non)-linear waves and currents. In Proceedings of the 24th International Conference on Coastal Engineering, Kobe, Japan, 23–28 October 1994. [CrossRef]
19. Fang, H.W.; Chen, M.H.; Chen, Q.H. One-dimensional numerical simulation of non-uniform sediment transport under unsteady flows. *Int. J. Sediment Res.* **2008**, *23*, 316–328. [CrossRef]
20. Caviedes-Voullieme, D.; Morales-Hernandez, M.; Juez, C.; Lacasta, A.; Garcia-Navarro, P. Two-Dimensional Numerical Simulation of Bed-Load Transport of a Finite-Depth Sediment Layer: Applications to Channel Flushing. *J. Hydraul. Eng.* **2017**, *143*. [CrossRef]
21. Soares-Frazao, S.; Zech, Y. HLLC scheme with novel wave-speed estimators appropriate for two-dimensional shallow-water flow on erodible bed. *Int. J. Numer. Methods Fluids* **2011**, *66*, 1019–1036. [CrossRef]
22. Gunsolus, E.H.; Binns, A.D. Effect of morphologic and hydraulic factors on hysteresis of sediment transport rates in alluvial streams. *River Res. Appl.* **2018**, *34*, 183–192. [CrossRef]

23. Bombar, G.; Elci, S.; Tayfur, G.; Guney, S.; Bor, A. Experimental and Numerical Investigation of Bed-Load Transport under Unsteady Flows. *J. Hydraul. Eng.* **2011**, *137*, 1276–1282. [CrossRef]

24. Mao, L. The effect of hydrographs on bed load transport and bed sediment spatial arrangement. *J. Geophys. Res. Earth Surf.* **2012**, *117*. [CrossRef]

25. Mao, L. The effects of flood history on sediment transport in gravel-bed rivers. *Geomorphology* **2018**, *322*, 196–205. [CrossRef]

26. Perret, E.; Berni, C.; Camenen, B.; Herrero, A.; Abderrezzak, K.E. Transport of moderately sorted gravel at low bed shear stresses: The role of fine sediment infiltration. *Earth Surf. Process. Landf.* **2018**, *43*, 1416–1430. [CrossRef]

27. Humphries, R.; Venditti, J.G.; Sklar, L.S.; Wooster, J.K. Experimental evidence for the effect of hydrographs on sediment pulse dynamics in gravel-bedded rivers. *Water Resour. Res.* **2012**, *48*. [CrossRef]

28. Wang, L.; Cuthbertson, A.J.S.; Pender, G.; Cao, Z. Experimental investigations of graded sediment transport under unsteady flow hydrographs. *Int. J. Sediment Res.* **2015**, *30*, 306–320. [CrossRef]

29. Li, Z.J.; Qian, H.L.; Cao, Z.X.; Liu, H.H.; Pender, G.; Hu, P.H. Enhanced bed load sediment transport by unsteady flows in a degrading channel. *Int. J. Sediment Res.* **2018**, *33*, 327–339. [CrossRef]

30. Redolfi, M.; Bertoldi, W.; Tubino, M.; Welber, M. Bed Load Variability and Morphology of Gravel Bed Rivers Subject to Unsteady Flow: A Laboratory Investigation. *Water Resour. Res.* **2018**, *54*, 842–862. [CrossRef]

31. Ferrer-Boix, C.; Hassan, M.A. Channel adjustments to a succession of water pulses in gravel bed rivers. *Water Resour. Res.* **2015**, *51*, 8773–8790. [CrossRef]

32. Waters, K.A.; Curran, J.C. Linking bed morphology changes of two sediment mixtures to sediment transport predictions in unsteady flows. *Water Resour. Res.* **2015**, *51*, 2724–2741. [CrossRef]

33. Wong, M.; Parker, G. One-dimensional modeling of bed evolution in a gravel bed river subject to a cycled flood hydrograph. *J. Geophys. Res. Earth Surf.* **2006**, *111*. [CrossRef]

34. Mrokowska, M.M.; Rowinski, P.M.; Ksiazek, L.; Struzynski, A.; Wyrebek, M.; Radecki-Pawlik, A. Laboratory studies on bedload transport under unsteady flow conditions. *J. Hydrol. Hydromech.* **2018**, *66*, 23–31. [CrossRef]

35. Curran, J.C.; Waters, K.A. The importance of bed sediment sand content for the structure of a static armor layer in a gravel bed river. *J. Geophys. Res. Earth Surface* **2014**, *119*, 1484–1497. [CrossRef]

36. Orru, C.; Blom, A.; Uijttewaal, W.S.J. Armor breakup and reformation in a degradational laboratory experiment. *Earth Surf. Dyn.* **2016**, *4*, 461–470. [CrossRef]

37. Curran, J.C.; Waters, K.A.; Cannatelli, K.M. Real time measurements of sediment transport and bed morphology during channel altering flow and sediment transport events. *Geomorphology* **2015**, *244*, 169–179. [CrossRef]

38. Martin, R.L.; Jerolmack, D.J. Origin of hysteresis in bed form response to unsteady flows. *Water Resour. Res.* **2013**, *49*, 1314–1333. [CrossRef]

39. Mrokowska, M.; Rowiński, P.; Książek, L.; Strużyński, A.; Wyrębek, M.; Radecki-Pawlik, A. Flume experiments on gravel bed load transport in unsteady flow—Preliminary results. In *Hydrodynamic and Mass Transport at Freshwater Aquatic Interfaces*; Rowiński, P., Ed.; Springer International Publishing Switzerland: Cham, Switzerland, 2016; pp. 221–233.

40. Nelson, J.M.; Logan, B.L.; Kinzel, P.J.; Shimizu, Y.; Giri, S.; Shreve, R.L.; McLean, S.R. Bedform response to flow variability. *Earth Surf. Process. Landf.* **2011**, *36*, 1938–1947. [CrossRef]

41. Piedra, M.M.; Haynes, H.; Hoey, T.B. The spatial distribution of coarse surface grains and the stability of gravel river beds. *Sedimentology* **2012**, *59*, 1014–1029. [CrossRef]

42. Shvidchenko, A.B.; Kopaliani, Z.D. Hydraulic modeling of bed load transport in gravel-bed Laba River. *J. Hydraul. Eng.* **1998**, *124*, 778–785. [CrossRef]

43. Ettema, R. Hydraulic modelling: Concepts and practice. In *ASCE Manuals and Reports on Engineering Practice No. 97*; American Society of Civil Engineers: Reston, VA, USA, 2000.

44. Carbonneau, P.E.; Bergeron, N.E. The effect of bedload transport on mean and turbulent flow properties. *Geomorphology* **2000**, *35*, 267–278. [CrossRef]

45. Nelson, J.M.; Shreve, R.L.; McLean, S.R.; Drake, T.G. Role of near-bed turbulence structure in bed-load transport and bed form mechanics. *Water Resour. Res.* **1995**, *31*, 2071–2086. [CrossRef]

46. Dixit, S.; Patel, P. Stochastic nature of turbulence over mobile bed channels. *J. Hydraul. Eng.* **2018**. [CrossRef]

47. Dey, S.; Das, R.; Gaudio, R.; Bose, S.K. Turbulence in mobile-bed streams. *Acta Geophys.* **2012**, *60*, 1547–1588. [CrossRef]
48. Bialik, R.J.; Nikora, V.I.; Karpinski, M.; Rowinski, P.M. Diffusion of bedload particles in open-channel flows: Distribution of travel times and second-order statistics of particle trajectories. *Environ. Fluid Mech.* **2015**, *15*, 1281–1292. [CrossRef]
49. Spiller, S.M.; Ruther, N.; Friedrich, H. Dynamic Lift on an Artificial Static Armor Layer during Highly Unsteady Open Channel Flow. *Water* **2015**, *7*, 4951–4970. [CrossRef]
50. Paiement-Paradis, G.; Marquis, G.; Roy, A. Effects of turbulence on the transport of individual particles as bedload in a gravel-bed river. *Earth Surf. Process. Landf.* **2011**, *36*, 107–116. [CrossRef]
51. Song, T.; Chiew, Y.M.; Chin, C.O. Effect of bed-load movement on flow friction factor. *J. Hydraul. Eng.* **1998**, *124*, 165–175. [CrossRef]
52. Hohermuth, B.; Weitbrecht, V. Influence of Bed-Load Transport on Flow Resistance of Step-Pool Channels. *Water Resour. Res.* **2018**, *54*, 5567–5583. [CrossRef]
53. Campbell, L.; McEwan, I.; Nikora, V.; Pokrajac, D.; Gallagher, M.; Manes, C. Bed-load effects on hydrodynamics of rough-bed open-channel flows. *J. Hydraul. Eng.* **2005**, *131*, 576–585. [CrossRef]
54. Recking, A.; Frey, P.; Paquier, A.; Belleudy, P.; Champagne, J.Y. Feedback between bed load transport and flow resistance in gravel and cobble bed rivers. *Water Resour. Res.* **2008**, *44*. [CrossRef]
55. Ghimire, B.; Deng, Z. Event Flow Hydrograph-Based Method for Modeling Sediment Transport. *J. Hydrol. Eng.* **2013**, *18*, 919–928. [CrossRef]
56. Graf, W.H.; Song, T. Bed-shear stresses in nonuniform and unsteady open-channel flows. *J. Hydraul. Res.* **1995**, *33*, 699–704. [CrossRef]
57. Mrokowska, M.M.; Rowinski, P.M.; Kalinowska, M.B. A methodological approach of estimating resistance to flow under unsteady flow conditions. *Hydrol. Earth Syst. Sci.* **2015**, *19*, 4041–4053. [CrossRef]
58. Rowinski, P.M.; Czernuszenko, W.; Pretre, J.M. Time-dependent shear velocities in channel routing. *Hydrol. Sci. J.* **2000**, *45*, 881–895. [CrossRef]
59. Aberle, J.; Rowiński, P.M.; Henry, P.Y.; Detert, M. Auxiliary hydrodynamic variables. Bed shear stress. In *Experimental Hydraulics, Volume 2: Methods, Instrumentation, Data Processing and Management*; Aberle, J., Rennie, C., Admiraal, D., Muste, M., Eds.; CRC Press: Boca Raton, FL, USA, 2017; pp. 322–332.
60. Tu, H.Z.; Graf, W.H. Friction in unsteady open-channel flow over gravel beds. *J. Hydraul. Res.* **1993**, *31*, 99–110. [CrossRef]
61. Song, T.; Graf, W.H. Velocity and turbulence distribution in unsteady open-channel flows. *J. Hydraul. Eng.* **1996**, *122*, 141–154. [CrossRef]
62. Nezu, I.; Nakagawa, H. Turbulence measurements in unsteady free-surface flows. *Flow Meas. Instrum.* **1995**, *6*, 49–59. [CrossRef]
63. Ferreira, R.M.L.; Franca, M.J.; Leal, J.G.A.B.; Cardoso, A.H. Flow over rough mobile beds: Friction factor and vertical distribution of the longitudinal mean velocity. *Water Resour. Res.* **2012**, *48*. [CrossRef]
64. Nikora, V.; McEwan, I.; McLean, S.; Coleman, S.; Pokrajac, D.; Walters, R. Double-averaging concept for rough-bed open-channel and overland flows: Theoretical background. *J. Hydraul. Eng.* **2007**, *133*, 873–883. [CrossRef]
65. Mrokowska, M.M.; Rowinski, P.M.; Kalinowska, M.B. Evaluation of friction velocity in unsteady flow experiments. *J. Hydraul. Res.* **2015**, *53*, 659–669. [CrossRef]
66. Bombar, G. Hysteresis and Shear Velocity in Unsteady Flows. *J. Appl. Fluid Mech.* **2016**, *9*, 839–853. [CrossRef]
67. Ghimire, B.; Deng, Z.-Q. Event flow hydrograph-based method for shear velocity estimation. *J. Hydraul. Res.* **2011**, *49*, 272–275. [CrossRef]
68. Meirovich, L.; Laronne, J.B.; Reid, I. The variation of water-surface slope and its significance for bedload transport during floods in gravel-bed streams. *J. Hydraul. Res.* **1998**, *36*, 147–157. [CrossRef]
69. Frey, P.; Church, M. Bedload: A granular phenomenon. *Earth Surf. Process. Landf.* **2011**, *36*, 58–69. [CrossRef]
70. Ferdowsi, B.; Ortiz, C.P.; Houssais, M.; Jerolmack, D.J. River-bed armouring as a granular segregation phenomenon. *Nat. Commun.* **2017**, *8*. [CrossRef]
71. Reid, I.; Frostick, L.E.; Layman, J.T. The incidence and nature of bedload transport during flood flows in coarse-grained alluvial channels. *Earth Surf. Process. Landf.* **1985**, *10*, 33–44. [CrossRef]
72. Bertin, S.; Friedrich, H. Effect of surface texture and structure on the development of stable fluvial armors. *Geomorphology* **2018**, *306*, 64–79. [CrossRef]

73. Haschenburger, J.K.; Wilcock, P.R. Partial transport in a natural gravel bed channel. *Water Resour. Res.* **2003**, *39*. [CrossRef]

74. Vericat, D.; Batalla, R.J.; Garcia, C. Breakup and reestablishment of the armour layer in a large gravel-bed river below dams: The lower Ebro. *Geomorphology* **2006**, *76*, 122–136. [CrossRef]

75. Clayton, J.A.; Pitlick, J. Persistence of the surface texture of a gravel-bed river during a large flood. *Earth Surf. Process. Landf.* **2008**, *33*, 661–673. [CrossRef]

76. Church, M.; Hassan, M.A. Mobility of bed material in Harris Creek. *Water Resour. Res.* **2002**, *38*. [CrossRef]

77. Venditti, J.G.; Dietrich, W.E.; Nelson, P.A.; Wydzga, M.A.; Fadde, J.; Sklar, L. Mobilization of coarse surface layers in gravel-bedded rivers by finer gravel bed load. *Water Resour. Res.* **2010**, *46*. [CrossRef]

78. Wilcock, P.R.; Kenworthy, S.T.; Crowe, J.C. Experimental study of the transport of mixed sand and gravel. *Water Resour. Res.* **2001**, *37*, 3349–3358. [CrossRef]

79. Curran, J.C. The decrease in shear stress and increase in transport rates subsequent to an increase in sand supply to a gravel-bed channel. *Sediment. Geol.* **2007**, *202*, 572–580. [CrossRef]

80. Church, M.; Ferguson, R.I. Morphodynamics: Rivers beyond steady state. *Water Resour. Res.* **2015**, *51*, 1883–1897. [CrossRef]

81. Julien, P.Y.; Klaassen, G.J.; Ten Brinke, W.B.M.; Wilbers, A.W.E. Case study: Bed resistance of Rhine River during 1998 flood. *J. Hydraul. Eng.* **2002**, *128*, 1042–1050. [CrossRef]

82. Reesink, A.J.H.; Parsons, D.R.; Ashworth, P.J.; Best, J.L.C.; Hardy, R.J.; Murphy, B.J.; McLelland, S.J.; Unsworth, C. The adaptation of dunes to changes in river flow. *Earth-Sci. Rev.* **2018**, *185*, 1065–1087. [CrossRef]

83. Bloschl, G.; Hall, J.; Parajka, J.; Perdigao, R.A.P.; Merz, B.; Arheimer, B.; Aronica, G.T.; Bilibashi, A.; Bonacci, O.; Borga, M.; et al. Changing climate shifts timing of European floods. *Science* **2017**, *357*, 588–590. [CrossRef]

84. Baynes, E.R.C.; van de Lageweg, W.I.; McLelland, S.J.; Parsons, D.R.; Aberle, J.; Dijkstra, J.; Henry, P.Y.; Rice, S.P.; Thom, M.; Moulin, F. Beyond equilibrium: Re-evaluating physical modelling of fluvial systems to represent climate changes. *Earth-Sci. Rev.* **2018**, *181*, 82–97. [CrossRef]

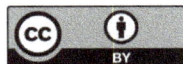

water

MDPI

Review

Water-Worked Gravel Bed: State-of-the-Art Review

Ellora Padhi [1], Subhasish Dey [1,*], Venkappayya R. Desai [1], Nadia Penna [2] and Roberto Gaudio [2]

[1]   Department of Civil Engineering, Indian Institute of Technology Kharagpur, West Bengal 721302, India;
      ellora@iitkgp.ac.in (E.P.); venkapd@civil.iitkgp.ac.in (V.R.D.)
[2]   Dipartimento di Ingegneria Civile, Università della Calabria, 87036 Rende, Italy;
      nadia.penna@unical.it (N.P.); gaudio@unical.it (R.G.)
*   Correspondence: sdey@iitkgp.ac.in; Tel.: +91-943-471-3850

Received: 6 March 2019; Accepted: 2 April 2019; Published: 4 April 2019

Abstract: In a natural gravel-bed stream, the bed that has an organized roughness structure created by the streamflow is called the water-worked gravel bed (WGB). Such a bed is entirely different from that created in a laboratory by depositing and spreading gravels in the experimental flume, called the screeded gravel bed (SGB). In this paper, a review on the state-of-the-art research on WGBs is presented, highlighting the role of water-work in determining the bed topographical structures and the turbulence characteristics in the flow. In doing so, various methods used to analyze the bed topographical structures are described. Besides, the effects of the water-work on the turbulent flow characteristics, such as streamwise velocity, Reynolds and form-induced stresses, conditional turbulent events and secondary currents in WGBs are discussed. Further, the results form WGBs and SGBs are compared critically. The comparative study infers that a WGB exhibits a higher roughness than an SGB. Consequently, the former has a higher magnitude of turbulence parameters than the latter. Finally, as a future scope of research, laboratory experiments should be conducted in WGBs rather than in SGBs to have an appropriate representation of the flow field close to a natural stream.

Keywords: fluvial hydraulics; gravel-bed stream; turbulent flow; water-worked gravel bed

1. Introduction

The topic of natural gravel-bed streams remains a continued research interest for several decades owing to its practical importance. The multifaceted fluid–particle interfaces yield spatial flow heterogeneities, in addition to temporal intermittencies, especially in the near-bed flow zone. It is therefore imperative to comprehend the turbulent flow physiognomies that arise from these complex turbulence mechanisms in natural streams to accurately estimate the resistance to flow and/or sediment transport rate. In fact, the bed surface topography in a natural gravel-bed stream possesses a spatially multifaceted, but coherently organized bed structure, because it is created by the natural erosion and deposition processes governed by continual flood cycles. In this process, a water-worked gravel bed (WGB) is developed in a natural gravel-bed stream (Figure 1). Specifically, water-worked refers to the work done by the flowing water on the sediment bed, for which the transport of sediments occurs in a natural stream. Hence, the gravel-bed produced by the action of flowing water is called a WGB. In contrast, in laboratory experimental studies, manmade gravel-bed is prepared by depositing and spreading gravels into the flume for a given thickness. Such a bed is called a screeded gravel bed (SGB) (Figure 2). The bed surface topography for such a bed is unorganized and randomly poised, being different from that of a WGB, even for a given identical gravel size distribution in both beds. Nonetheless, in the laboratory, a WGB can be produced if an SGB is water-worked, which can mobilize the surface gravels over a sufficiently long period until the transport of gravels ceases. To the best of the authors' knowledge, a review on the studies done on WGBs has not yet been compiled, although there exist several compilations on the studies done on SGBs.

Figure 1. Photograph of a natural gravel-bed stream with a flow direction right to left (**a**); and close view of the natural gravel-bed stream with a flow direction left to right (**b**).

Figure 2. Photograph of an SGB in a laboratory flume (**a**); and a close view (**b**).

The aim of this article is therefore to compile a comprehensive state-of-the-art review of the most important laboratory experimental research on WGBs by analyzing the results scientifically. Attention is primarily paid to the WGB topography structures, time-averaged flow field, turbulent stresses, conditional turbulent events and secondary currents. In some cases, the results in WGBs are compared with those in SGBs.

2. Analysis of Bed Roughness Structures in WGBs

As an earlier practice, river engineers used the particle-size distribution curve to characterize the bed roughness. Bathurst [1] proposed the Nikuradse's equivalent roughness k_s approximately to be 3–3.5 d_{84}, where d_{84} is the 84th percentile of a particle-size distribution belonging to the range of 240–500 mm. However, the roughness height k_s cannot be estimated considering only a single gravel size, e.g., d_{84}, because other factors, such as gravel shape, orientation, alignment and structural arrangements, are of equal importance, providing significant impact on the estimation of the roughness height k_s [2–7]. Considering this fact, Kirchner et al. [5] were the first to measure the friction angle of the sediment mixtures having median size d_{50} ranging 1.2–12 mm in both WGBs and SGBs. They observed that the difference in friction angles increases with a decrease in gravel size. Tp To quantify the impact of water-work on the bed roughness structures, they compared the difference between the bed roughness structures in WGBs that were created by the water action and SGBs that were manmade. The results reveal that, for an identical particle-size distribution, the distributions of friction angles in a WGB and an SGB are different, because the friction angles in the former are smaller than those in the latter. They argued that the difference in the friction angles occurs owing to the difference in gravel packing geometries in these beds. Hence, they concluded that the friction angle measured in an SGB cannot be directly applicable to a WGB.

2.1. Use of Probability Density Function

Skewness and kurtosis coefficients are deemed to be important properties, as they provide useful information about the distributions of roughness structures. Kirchner et al. [5], Nikora et al. [8], Marion et al. [9] and Aberle and Nikora [10] performed preliminary analysis of the roughness structures in WGBs and SGBs. They found that the probability distribution functions (PDFs) of roughness structures in both beds follow the normal distribution, but possess skewness values of opposite signs. The PDFs in WGBs are positively skewed, while they are negatively skewed in SGBs. Marion et al. [9] analyzed the PDFs of roughness structures at different time intervals during the creation of a WGB by the sediment transport process. The PDFs of roughness structures were narrow at the initial periods, but, as time elapsed, they became flatter having increased skewness. Although in their study the PDFs of different roughness structures were unable to identify the particle scale features uniquely, the WGB PDFs are positively skewed and the SGB PDFs are negatively skewed. Aberle et al. [10] also had similar observations. They postulated that, in a WGB, the roughness structure is composed of course gravels with finer particles filling the interstices of gravels. This causes the reduction of the surface elevation of the roughness structure with respect to the mean bed level. As a result, the PDF is positively skewed. Buffin-Bélanger et al. [11] also reported similar results. Later et al. [12] performed experiments in WGBs and an SGB. In their study, the WGBs were created by feeding sediment at the upstream end of the flume. Besides, to understand the impact of feeding rate on the roughness structures, they generated three WGBs at three different feeding rates. These WGBs, namely Fed bed 1, Fed bed 2 and Fed bed 3, were created at feeding rates of 0.0624 kg m^{-1}, 0.0922 kg m^{-1} and 0.152 kg m^{-1}, respectively. The PDFs of these four beds revealed that the SGB has a slightly negatively skewed distribution of the roughness structure, while the WGBs have a very slightly positively skewed distribution (Figure 3). Moreover, they found that, with an increase in feeding rate, the skewness value decreases. Interestingly, in both the WGBs and the SGB, the kurtosis coefficients were found to be positive, indicating a leptokurtic curve.

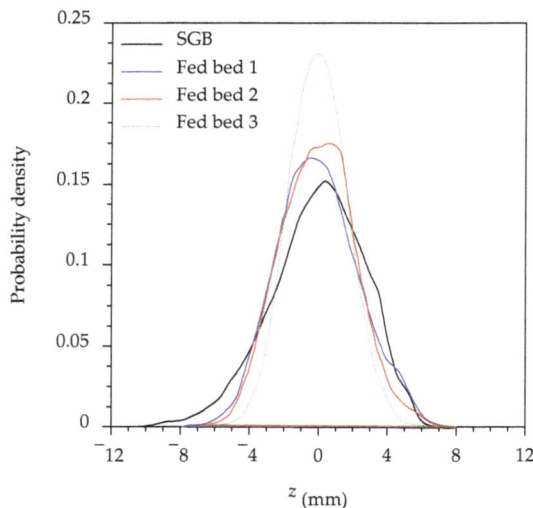

Figure 3. Probability density functions of bed surface fluctuations with respect to the mean surface level in WGBs (namely, Fed bed 1, Fed bed 2, and Fed bed 3) and an SGB (data extracted from Cooper and Tait [12]).

2.2. Use of the Second-Order Structure Function

Although the PDFs of the roughness structures provide information about the bed surface features and the difference between a WGB and an SGB, they do not provide any information about the degree of organization of the surface particles. For an immobile gravel-bed with a Gaussian distribution of roughness structure, a quantitative evaluation of roughness structure can be done by using a second-order structure function. Nikora et al. [8] were the pioneers of applying the second-order structure function to ascertain the behavioral features in a WGB. They argued that, if a roughness structure is treated as a random field, then the roughness description can be reduced to two-dimensional spectrum, correlation function and structure function, by using the hypothesis of local spatial homogeneity. The advantage of using the second-order structure function is that it can consider all the important parameters that influence the roughness structure. The second-order structure function is expressed as follows:

$$D(l_x, l_y) = \frac{1}{(N-n)(M-m)} \sum_{j=1}^{N-n} \sum_{k=1}^{M-m} \left[\left| z'(x_j + n\delta x, y_k + m\delta y) - z'(x_j, y_k) \right| \right]^2, \tag{1}$$

where $D(l_x, l_y)$ is the second-order structure function of the bed elevation, z' is the bed surface fluctuation with respect to the mean bed elevation $\bar{z}$, l_x is the sampling interval in streamwise direction ($= n\delta x$), l_y is the sampling interval in spanwise direction ($= m\delta y$), $j = 1, 2, 3, \ldots, n$, n is the number of points in streamwise direction, $k = 1, 2, 3, \ldots, m$, m is the number of points in spanwise direction, and δx and δy are the sampling intervals in x and y directions, respectively. However, following the concept of Monin et al. [13], Nikora et al. [8] obtained the $D(l_x, l_y)$ as follows:

$$D(l_x, l_y) = 2(\sigma_z^2 - R(l_x, l_y)), \tag{2}$$

where

$$\sigma_z = \sqrt{\frac{1}{NM-1} \sum_{i=1}^{NM} z'^2_i} \text{ and}$$

$$R(l_x, l_y) = \frac{1}{(N-n)(M-m)} \sum_{j=1}^{N-n} \sum_{k=1}^{M-m} [z'(x_j + n\delta x, y_k + m\delta y) z'(x_j, y_k)]. \tag{3}$$

In Equation (3), σ_z is the standard deviation and $R(l_x, l_y)$ is the correlation function of the bed elevations. Interestingly, Nikora et al. [8], instead of computing $D(l_x, l_y)$ for all the measured points, computed the $D(l_x, l_y = 0)$ and $D(l_x = 0, l_y)$ assuming that the main anisotropy axes of the roughness coincide with their chosen x and y axes in a WGB and an SGB. From the analysis, they found that the data form both beds collapsed onto two curves, indicating the existence of two different universal classes of gravel-bed roughness.

Further, Nikora et al. [8] observed that the $D(l_x, l_y)$ is composed of three regions: scaling, transition and saturation regions. However, the first and last regions play the main role. For a small spatial lag, the $D(l_x, l_y)$ acts as a power function (that is, the scaling region), while, for a large spatial lag, the $D(l_x, l_y)$ becomes constant (that is, the saturation region). Goring et al. [14] extended the work of Nikora et al. [8] by computing the second-order structure function for a two-dimensional roughness structure and obtained similar results. Later, Butler et al. [15] analyzed the roughness structure in a WGB using the fractal analysis in both streamwise and spanwise directions. To do so, they applied a two-dimensional fractal method to high-resolution digital elevation models. They identified a mixed fractal behavior with two characteristic fractal bands; one associated with the subgrain scale and the other associated with the grain scale. The subgrain and grain scales features are isotropic and anisotropic, respectively. They also observed that, owing to the streamwise orientation of the longest axis of particles, the fractal dimensions are higher in the streamwise direction than in the other directions. It implies that the effects of water-work are to modify the organization of roughness structure by increasing the surface irregularities and hence the roughness height. Then, similar to Goring et al. [14], Marion et al. [9] used Equation (2) to analyze the variation of roughness structure with time in a WGB under a mobile bed condition. They showed that the roughness structures captured at different time intervals are directly associated with the bed mobility conditions. Moreover, their second-order structure function for roughness structure depicted the development of two different grain scale classes. One grain class was developed under a static armoring condition, where the gravels formed a bed surface with strong streamwise and spanwise coherences. The other one was developed under the dynamic armoring condition, where the gravels formed a bed surface very quickly, but only with very strong streamwise coherence. However, they were unable to establish a relation between the grain scale features with the bed mobility condition. Following the method proposed by Nikora et al. [8], Cooper and Tait [12] used the second-order structure functions for roughness structures of three fed WGBs and an SGB. They calculated the correlation lengths in the WGBs and an SGB. They indicated that the correlation lengths in both streamwise and spanwise directions in the WGBs are larger than those in the SGB, confirming that the WGBs have larger scale bed features than the SGB. Later, Qin and Ng [16] performed the second-order structure function analysis in a WGB and an SGB. They found similar results as obtained by the aforementioned researchers.

2.3. Use of Higher-Order Structure Function

Owing to gravel imbrications (overlapping of gravels) and orientations, it is not always feasible to obtain a PDF of Gaussian distribution. Hence, for such a case, to quantify the features of the roughness structures in a WGB, the higher-order structure function is deemed to be an effective tool [8,10,17]. Further, the use of higher-order structure function provides multiscaling behavioral features of the roughness structure [18,19]. It is represented as follows:

$$D_p(l_x, l_y) = \frac{1}{(N-n)(M-m)} \sum_{j=1}^{N-n} \sum_{k=1}^{M-m} \left[|z'(x_j + n\delta x, y_k + m\delta y) - z'(x_j, y_k)| \right]^p \qquad (4)$$

where $D_p(l_x, l_y)$ is the higher-order structure function and p is the order of the moment of the structure function.

Nikora et al. [17] computed the structure function up to the sixth-order ($p = 6$) using the concept of Kolmogorov [20]. Similar to the second-order structure function, the shape of the $D_p(l_x, l_y)$ can be divided into three regions: scaling, transition and saturation regions (Figure 4) [17]. Interestingly, they revealed that, in WGBs, the boundary between the scaling and transition regions is of the order of the median gravel size d_{50}, while the boundary between the transition and saturation regions is of order of d_{90}, where d_{90} is the 90th percentile of a particle-size distribution. Akin to Nikora et al. [8], in Nikora and Walsh [17], the analysis of $D_p(l_x, l_y)$ suggested that the grain scales are isotropic, indicating that the $D_p(l_x, l_y)$ is independent of the axis rotation. However, within the transition region, the $D_p(l_x, l_y)$ becomes anisotropic. Further, within the scaling region, the scaling exponent ξ plays an important role. For small values of l_x and l_y, the ξ varies linearly with p. Nevertheless, as p increases, the ξ varies nonlinearly with p, suggesting a multiscaling behavior of roughness structures in WGBs, being sensitive to the flow direction (Figure 5). The reason is attributed to the shape and the spatial arrangements of gravels [17].

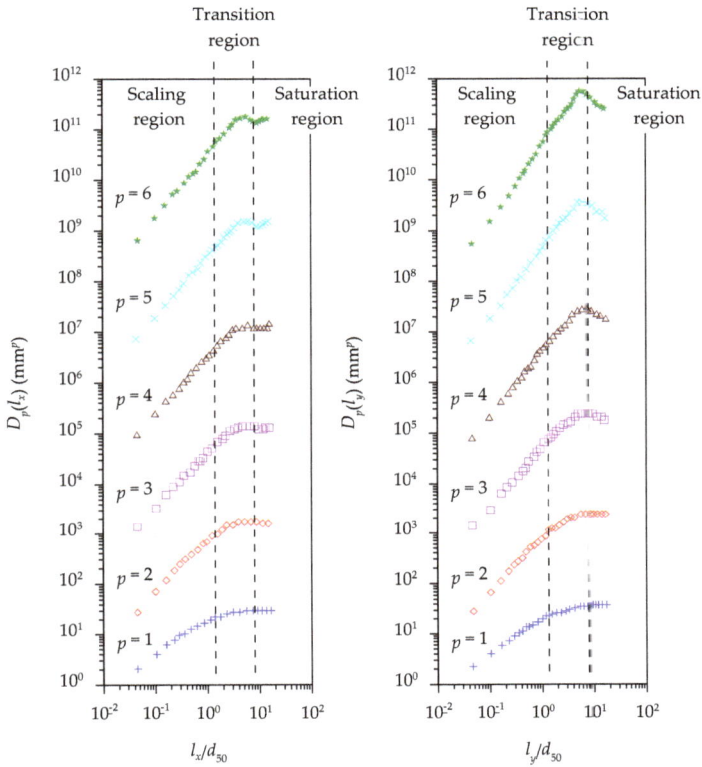

Figure 4. Non-dimensional structure function of a WGB (data extracted from Nikora and Walsh [17]).

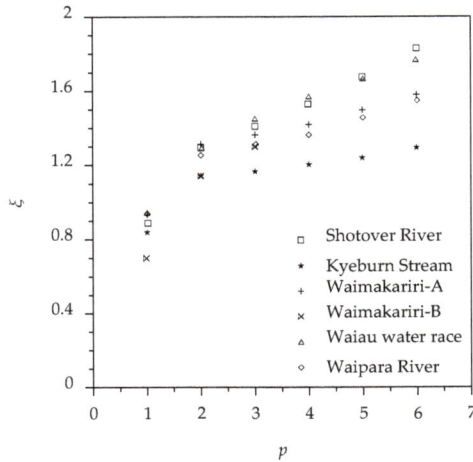

Figure 5. Variations of scaling exponents ξ of the generalized structure function with the order p (data extracted from Nikora and Walsh [17]).

In a WGB, to understand the effects of multiscaling behavior of roughness structures in the flow direction, Aberle et al. [10] used both the second- and higher-order structure functions. They plotted the second-order structure functions in the form of contours for different armoring discharge conditions. In all cases, the contours form an elliptical shape. Similar observations were also made by Butler et al. [15] and Nikora et al. [17]. An examination of the contours reveals that, at small spatial lags in both streamwise and vertical directions, the longest axes of the nearly elliptical contours are aligned in the streamwise direction (Figure 6). It implies that the majority of gravels are to be rested on the bed keeping their longest axis in the streamwise direction [10]. They further argued that at the end of the armoring process, the gravels keep their longest axis in the streamwise direction before coming to the resting position, which is in conformity with the observations of Allen [21]. On the other hand, large gravels, which were not moved by the flow, are oriented without any directional preference. As a result, the alignment of the large contours does not match with the flow direction (Figure 6). By analyzing the higher-order structure functions in WGBs for different armoring discharge conditions (0.012–0.025 $m^3 s^{-1}$), Aberle and Nikora [10] found that the WGB roughness structures possess multiscaling behavioral features, which is in conformity with the findings of Nikora et al. [17].

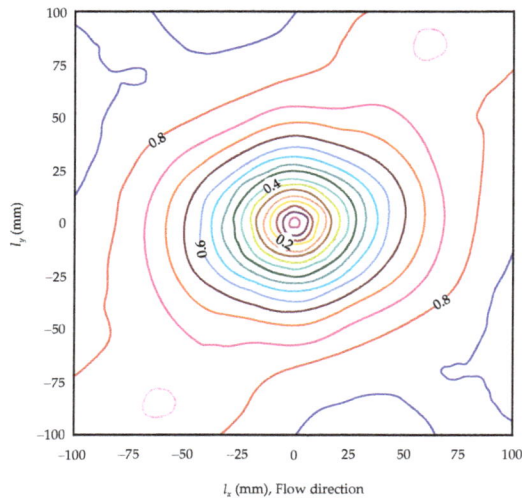

Figure 6. Contours of the second-order structure function for the roughness structure at a discharge of 0.25 $m^3 s^{-1}$ (data extracted from Aberle and Nikora [10]).

Moreover, Aberle et al. [10] stated that, although the second-order structure function shows the orientation toward the flow, it does not provide any information about the flow direction; for instance, whether it took place from right to left or left to right. Hence, they used the method proposed by Smart et al. [22] to predict the flow direction based on the gravel orientation. By using the local bed inclination at small spatial lags, the average positive and negative bed slopes can be estimated as follows:

For a positive slope $[z'(x_j + n\delta x) - z'(x_j) > 0]$,

$$E_p(l_x) = \frac{1}{n_p} \sum_{j=1}^{n_p} \left[\frac{z'(x_j + n\delta x) - z'(x_j)}{n\delta x} \right] \tag{5}$$

and for a negative slope $[z'(x_j + n\delta x) - z'(x_j) < 0]$,

$$E_n(l_x) = \frac{1}{n_n} \sum_{j=1}^{n_n} \left[\frac{z'(x_j + n\delta x) - z'(x_j)}{n\delta x} \right], \tag{6}$$

where E_p and E_n are the positive and negative bed slopes, respectively, and n_p and n_n are the number of positive and negative slopes, respectively.

Using Equations (5) and (6), they calculated the E_p and E_n in the WGEs and found that, at small lags, the frequency of negative slope is less than that of positive slope, owing to the gravel imbrication. Further, they revealed that the ξ of the gravels is directly associated with the armoring discharge and the individual large gravels create less complex roughness structure than a large number of small gravels.

Moreover, to determine the self-affine property and complexity of roughness structure in a WGB, estimation of the Hurst coefficient is required [8,10,12,14]. In fact, the Hurst coefficient is related to the fractal dimension. For lower values of Hurst coefficient, the fractal dimension becomes higher, and the number of significant modes that enter the evaluation of the distributed field of roughness structure also remain higher. Within the scaling region, if the Hurst coefficients in streamwise and spanwise directions are the same, then the roughness structures are isotropic within the scaling region and vice versa. Nikora et al. [8] and Cooper et al. [12] estimated the Hurst coefficients in WGBs and SGBs. They found that the Hurst coefficients in WGBs are higher than those in SGBs, indicating the roughness structures in WGBs are more complex than those in SGBs [23]. Aberle et al. [10] estimated the Hurst coefficients for armoring discharge conditions, finding that the Hurst coefficient increases with an increase in discharge and is highly dependent on the shape and orientation of gravels.

3. Turbulence Characteristics in WGBs

Flow over a gravel-bed is spatially heterogeneous and this affects the entire turbulence structure in the flow. The continuous fluid–particle interaction causes a more complex near-bed flow field, making difficult to estimate the sediment transport, the resistance to flow and important turbulence parameters in the flow. It implies that the bed topography is the primary cause to have such a complex flow field. Thus far, several studies show that the bed topography in a WGB is fairly different from that in an SGB, as mentioned in the preceding section, suggesting that the impacts of both the beds on the turbulence characteristics are different. Considering this, several researchers analyzed the effects of the WGB roughness structures on the turbulence parameters for various flow conditions.

3.1. Effects of Water-Work on Streamwise Velocity

Barison et al. [24] analyzed the time-averaged flow field over a WGB and found that the flow field is drastically affected by the roughness structure owing to the action of water-work. However, in their study, they did not analyze the bed topography precisely. Later, Buffin-Bélanger et al. [11] analyzed the spatial heterogeneity in the flow parameters, especially at the near-bed flow zone, considering three different Reynolds numbers (1.7×10^5, 2.2×10^5 and 2.9×10^5) in a WGB. They observed that the spatial heterogeneity of the time-averaged velocity decreases with a decrease in the vertical distance, but it increases with an increase in Reynolds number. At a high Reynolds number, the spatial heterogeneity was found to be maximum in the near-bed flow zone. Further, they analyzed the mean and skewness maps of the time-averaged streamwise velocity on the horizontal plane at two different vertical distances: one near the bed and the other in the main flow layer. They observed that the mean and skewness maps for the near-bed case were more complex than those for the main flow layer case. The skewness values suggested that the shapes of the velocity distributions are different for these cases. In the near-bed flow zone, the skewness is mostly positive, while in the main flow layer, the skewness is almost negative. Buffin-Bélanger et al. [11] argued that the positive skewness values in the near-bed flow zone possibly reflect incursions of high-speed fluid streaks, while the negative values in the main flow layer indicate the incursions of low-speed fluid streaks. To be explicit, the low-speed and high-speed fluid streaks refer to the ejections and sweeps.

To understand the effects of the bed roughness structure on the spatial organization of the flow structure, Cooper et al. [25] conducted experiments on two WGBs, where the bed structure was created using unimodal and bimodal gravel mixtures. They analyzed the streamwise velocities in both the beds and found that, although the bed roughness structures are different, the spatial organization of streamwise velocities in both the beds are almost the same. Further, Cooper et al. [25] also studied the effects of relative submergence (1.2–1.9 for unimodal gravel-bed and 1.3–2 for bimodal gravel-bed) on the spatial pattern of streamwise velocity showing them in the form of contour plots. They found that, with an increase in relative submergence, the number of high-speed fluid streaks decreases, but the number of low-speed fluid streaks increases. It implies that, as relative submergence increases, the streamwise velocity distribution becomes spatially homogeneous, which is in conformity with the observations of Legleiter et al. [26]. In both unimodal and bimodal gravel-beds, they also found that, for a given slope and bed shear stress, the relative submergence provides a more significant impact on the spatial distribution of the streamwise velocity than the bed topography. Later, Hardy et al. [27] performed a time series analysis to visualize the instantaneous velocity field through a series of consecutive images in WGBs, for three different Reynolds numbers (1.3×10^5, 2.5×10^5 and 2.7×10^5). They observed that, for all Reynolds numbers, the flows are highly inconsistent in the near-bed flow zone. Further, the turbulent structures that originate from the near-bed zone are to intrude into the main flow layer. These structures change their form and magnitude at higher Reynolds numbers, becoming more distinct, having a clearer velocity signature and a steeper upstream-dipping slope.

Thereafter, Koll et al. [28] studied the near-bed turbulent flow field over two WGBs. They kept the statistical distribution of the surface gravels identical in both the beds, but with different gravel orientations. In the first phase, they created a WGB and took the flow measurements over it. Subsequently, they rotated the surface gravels in a WGB by 90° and measured the flow field in the newly created WGB. Analysis of the double averaged (DA) streamwise velocity $\langle \bar{u} \rangle$ profiles in both the beds showed that a higher flow retardation occurs in the WGB with rotated gravels than in the original WGB. They identified that the difference in magnitude of $\langle \bar{u} \rangle$ is mainly caused by the change in near-bed turbulence rather than by the spatial distribution of time-averaged velocity.

Besides, after Nezu et al. [29], the bed topography can be considered as one of the most influencing factors in estimating the turbulence parameters. Therefore, to quantify the impact of the bed topography on the flow velocity, Pu et al. [30] carried out experiments over three different beds (a smooth bed, a WGB and an SGB), using an Acoustic Doppler Velocimeter (ADV), and compared the results. They used the following equations of log-wake laws for velocity profiles:

For smooth flow,

$$\frac{u}{u_*} = \frac{1}{\kappa} \ln\left(\frac{u_* z}{\nu}\right) + \underbrace{\frac{1}{\kappa} \ln\left(\frac{u_* z_0}{\nu}\right)}_{B_r} + \frac{2\Pi}{\kappa} \sin^2\left(\frac{\pi}{2}\frac{z}{h}\right), \tag{7}$$

and for rough (SGB and WGB) flows,

$$\frac{u}{u_*} = \frac{1}{\kappa} \ln\left(\frac{z + \Delta z}{k_s}\right) + \underbrace{\frac{1}{\kappa} \ln\left(\frac{z_0}{k_s}\right)}_{B_r} + \frac{2\Pi}{\kappa} \sin^2\left(\frac{\pi}{2}\frac{z}{h}\right), \tag{8}$$

where u_* is the frictional velocity, z is the vertical distance, ν is the kinematic viscosity, B_r is the constant of integration, κ is the von Kármán coefficient, Π is the Coles' wake parameter, Δz is the virtual bed level ($\approx 0.25\, k_s$, according to Dey et al. [31]), z_0 is the zero velocity level, k_s is the average roughness height, and h is the flow depth.

Pu et al. [30] used the velocity data of each bed to obtain the fitted curves for the log-wake laws (Figure 7). Interestingly, the values of B_r are lower in both the smooth and the rough (WGB and SGB) beds than the traditional values: $B_r = 5.5$ and 8.5 for the smooth and rough beds, respectively. Further, even though the flow conditions of both the WGB and SGB were identical, they observed that the B_r in the WGB was smaller than that in the SGB. It implies that, in the near-bed flow zone, a WGB roughness structure affects B_r and, in turn, the velocity profile. In addition, the comparison of Π values in the WGB and SGB revealed that the values of Π remain the same in the velocity profiles of both the beds, suggesting that the water-work has an insignificant impact on the Π, which mainly governs the velocity profile in the outer layer.

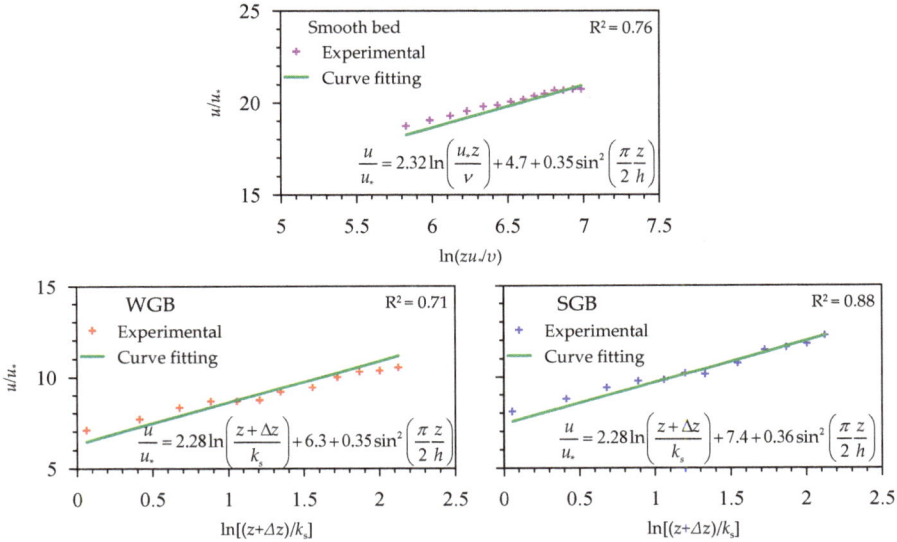

Figure 7. Variations of non-dimensional streamwise velocity with non-dimensional vertical distance zu_*/ν and $(z + \Delta z)/k_s$ in smooth and rough beds (WGB and SGB), respectively (data extracted from Pu et al. [30]).

To ascertain the impact of the water-work on streamwise velocity more precisely, using a unimodal sediment mixture, Padhi et al. [32] measured streamwise velocity in a WGB using a Particle Image Velocimetry (PIV) system and compared it with that in an SGB (Figure 8). In their study, owing to the water action, the WGB roughness structure was observed to be better organized than the SGB roughness structure, where gravels were randomly sorted. Akin to other rough-flow, in the study by Padhi et al. [32], owing to the presence of gravels, the values of $\langle \bar{u} \rangle$ in both the WGB and SGB are small in the near-bed flow zone. However, they gradually increase with an increase in vertical distance, reaching a maximum at the free surface. Moreover, Padhi et al. [32] found that, close to the bed, the $\langle \bar{u} \rangle$ in the WGB is higher than that in the SGB, although the flow conditions in both the beds were alike. They argued that the well-organized roughness structure in a WGB makes the near-bed flow more streamlined than that in an SGB, inducing the $\langle \bar{u} \rangle$ to attain a higher magnitude in the former than in the latter. However, the difference in magnitudes of $\langle \bar{u} \rangle$ between a WGB and an SGB gradually diminishes, as the vertical distance increases.

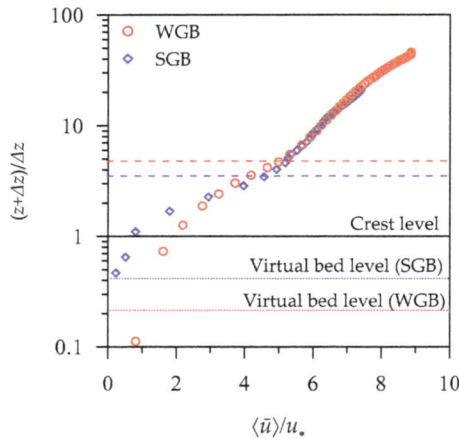

Figure 8. Variations of non-dimensional DA streamwise velocity with non-dimensional vertical distance $(z + \Delta z)/\Delta z$ in the WGB and SGB. The red and blue broken lines indicate the form-induced sublayers in the WGB and SGB, respectively (data extracted from Padhi et al. [32]).

3.2. Effects of Water-Work on Reynolds Shear Stresses and Form-Induced Shear Stresses

For steady, uniform flow over a macro-rough bed, the spatially averaged (SA) total fluid shear stress $\langle \overline{\tau} \rangle$ can be expressed as follows:

$$\langle \overline{\tau} \rangle = \left\langle \tau_f \right\rangle + \langle \overline{\tau}_{uw} \rangle + \langle \overline{\tau}_v \rangle, \tag{9}$$

where $\langle \tau_f \rangle$ is the SA form-induced shear stress $(= -\rho \langle \widetilde{u}\widetilde{w} \rangle)$, ρ is the mass density of fluid, $\widetilde{u}$ and $\widetilde{w}$ are the spatial velocity fluctuations in the streamwise and vertical directions, respectively, $\langle \overline{\tau}_{uw} \rangle$ is the SA Reynolds shear stress $(= -\rho \langle \overline{u'w'} \rangle)$, u' and w' are the temporal velocity fluctuations in the streamwise and vertical directions, respectively, and $\langle \overline{\tau}_v \rangle$ is the SA viscous shear stress $(= -\rho v \mathrm{d}\langle \overline{u} \rangle/\mathrm{d}z)$.

Although the $\langle \overline{\tau}_{uw} \rangle$ remains the prevailing stress in a turbulent flow across the flow depth, the $\langle \tau_f \rangle$ is the governing stress near the gravel-bed [33]. Aberle et al. [34] focused on the $\langle \tau_f \rangle$ profile influenced by the roughness elements. They analyzed the spatial flow heterogeneity in terms of the $\langle \tau_f \rangle$ in WGBs for different discharges. Their results infer that the magnitude of $\langle \tau_f \rangle$ is small away from the crest. However, the $\langle \tau_f \rangle$ increases as one moves toward the crest in the downward direction. It indicates that the $\widetilde{u}$ and $\widetilde{w}$ near the bed are higher than those away from the bed, resulting in a higher magnitude of $\langle \tau_f \rangle$. Further, they found that the $\langle \tau_f \rangle$ profiles are similar for different discharges. Interestingly, the similarity in $\langle \tau_f \rangle$ profiles is not preserved for different bed slopes. It suggests that for a given bed slope, the $\langle \tau_f \rangle$ profile is independent of discharge.

Then, to study the effects of different sediment mixtures on both the $\langle \overline{\tau}_{uw} \rangle$ and $\langle \tau_f \rangle$, Cooper and Tait [35] analyzed all the terms of Equation (9) in two WGBs created by the unimodal and bimodal sediment mixtures. For the unimodal sediment mixture, the relative submergences varied within the range of 1.2–1.9, while, for bimodal sediment mixture, they varied within 1.3–2. They analyzed the results in terms of forces caused by the shear stresses. In doing so, they considered the fluid force caused by the $\langle \overline{\tau}_{uw} \rangle$ at a given vertical distance as $\langle \overline{\tau}_{uw} \rangle \phi A_0$, where ϕ is the roughness geometric function $(= A_f/A_0$, where A_f is the area of fluid in the averaging domain at a given elevation within the total area $A_0)$. Above the roughness crest, $\phi = 1$. Similarly, the fluid force caused by the $\langle \tau_f \rangle$ at given vertical distance was obtained as $\langle \tau_f \rangle \phi A_0$. They further argued that in addition to these two forces, there exists an additional force called the form drag $\langle \tau_d \rangle$, which can be computed as $\int_z^{z_c} 0.5 C_d \rho \langle \overline{u} \rangle^2 A_e \mathrm{d}z$, where C_d is the drag coefficient and A_e is the exposed frontal area of the grain to the fluid. However, in the analysis, they neglected the force caused by the $\langle \overline{\tau}_v \rangle$ term considering that it has minimal impact on the turbulent flow. It is pertinent to mention that their analysis mainly focused on the zone below the roughness crest. Analyzing the forces, they inferred that within this zone, the vertical variations of the forces contributed from $\langle \overline{\tau}_{uw} \rangle$, $\langle \tau_f \rangle$ and $\langle \tau_d \rangle$ with ϕ are similar and thus controlled by the geometry of the roughness elements. Moreover, they observed that with a decrease in vertical distance, the reduction in the force caused by a damping of $\langle \overline{\tau}_{uw} \rangle$ is compensated by the addition of the force caused by $\langle \tau_d \rangle$. Furthermore, their results showed that as the relative submergence increases, the forces contributed from the $\langle \overline{\tau}_{uw} \rangle$, $\langle \tau_f \rangle$, and $\langle \tau_d \rangle$ increase. It suggests that for a given bed surface topography, the mechanism of momentum transfer between the fluid and particle fairly changes with an increase in relative submergence. When the results of unimodal and bimodal WGBs were compared, they found that for a given vertical distance and relative submergence, the force contributed from the $\langle \overline{\tau}_{uw} \rangle$ and $\langle \tau_d \rangle$ in the unimodal WGB is less than that in the bimodal WGB in the upper portion of the roughness layer and vice versa. Interestingly, in the lower portion of the roughness layers of both beds, the force caused by the $\langle \tau_f \rangle$ was observed to have different vertical distributions. It indicates that for a given relative submergence, the mechanism of momentum transfer differs owing to the difference in roughness structure.

Later, Cooper et al. [36] used the experimental data of Aberle et al. [54] to quantify the spatial flow variance and the $\langle \tau_f \rangle$ for different flow submergence conditions and for gravel-beds with different roughness structures. They observed that the spatial flow variance within the roughness layer is typically 4–5 times higher than that above the roughness layer. In fact, it becomes invariant to the vertical distance at a distance twice the roughness height above the crest. Owing to the increase in relative submergence, the spatial flow variance with respect to $\langle \tau_f \rangle$ decreases within and above the roughness layer. However, the flow submergence does not have a significant impact on the spatial flow variance with respect to $\langle \overline{\tau}_{uw} \rangle$. Further, their study infers that for different bed surface topographies, the spatial flow variance and the $\langle \tau_f \rangle$ profiles vary, suggesting that the bed geometry possesses a strong control on the spatial flow variance profiles and the vertical organization of the time-averaged flow within the roughness layer.

Pu et al. [30] compared the $\langle \overline{\tau}_{uw} \rangle$ profiles in a smooth bed with those in the WGB and SGB. They showed that the $\langle \overline{\tau}_{uw} \rangle$ profile in the smooth bed converges with the gravity line at a shorter vertical distance than those in the WGB and SGB. Between these two rough beds, the $\langle \overline{\tau}_{uw} \rangle$ profile in the WGB takes longer vertical distance to collapse on the gravity line than that in the SGB. They argued that as the WGB possesses higher roughness among all the beds, it causes the flow to have the thickest unsettled turbulence mixing layer in the near-bed flow zone, although the effects of roughness do not persist in the main flow layer. Further, regarding the magnitude of $\langle \overline{\tau}_{uw} \rangle$ profile, they showed that the $\langle \overline{\tau}_{uw} \rangle$ profile in a smooth bed attains the highest magnitude among all, but no explanation was given for that. Moreover, although they showed the effects of roughness on the $\langle \overline{\tau}_{uw} \rangle$, the impact of roughness on the $\langle \tau_f \rangle$ were not taken into consideration.

Recently, Padhi et al. [32] examined the effects of roughness on the $\langle \overline{\tau}_{uw} \rangle$ and $\langle \tau_f \rangle$ profiles in a WGB and an SGB. Akin to Pu et al. [30], Padhi et al. [32] found that the roughness height in the WGB was also higher than that in the SGB. However, the results of Padhi et al. [32] do not correspond to those of Pu et al. [30]. In the study by Padhi et al. [32], the $\langle \overline{\tau}_{uw} \rangle$ profile in the WGB is higher than that in the SGB owing to a higher roughness height in the former than in the latter (Figure 9). They stated that a higher roughness in the WGB than in the SGB enhances the u' and w' values, causing an increased magnitude of $\langle \overline{\tau}_{uw} \rangle$ in the WGB. The results are in agreement with those reported in Nezu and Nakagawa [29], Nikora et al. [33], Mignot et al. [37] and Dey and Das [38]. Moreover, the $\langle \overline{\tau}_{uw} \rangle$ profile in the WGB collapses on the gravity line at a shorter distance than that in the SGB. It implies that, although the WGB exhibits a higher roughness height than the SGB, owing to the well-organized roughness structure in the WGB, intense flow mixing is restricted to a shorter vertical distance. Further, the $\langle \tau_f \rangle$ profiles in the WGB and SGB showed that a higher roughness in the WGB than in the SGB produces large values of $\tilde{u}$ and $\tilde{w}$ causing an increased magnitude of $\langle \tau_f \rangle$ in the former than that in the latter (Figure 9). This suggests that in the near-bed flow zone, the flow is more heterogeneous in the WGB than in the SGB.

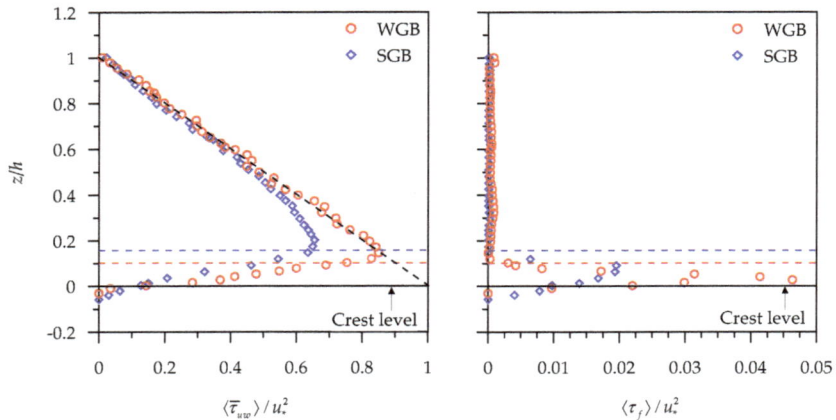

Figure 9. Variations of non-dimensional SA Reynolds shear stress $\langle \overline{\tau}_{uw} \rangle / u_*^2$ and SA form-induced shear stress $\langle \tau_f \rangle / u_*^2$ with non-dimensional vertical distance z/h in the WGB and SGB (data extracted from Padhi et al. [32]).

3.3. Effects of Water-Work on Reynolds Normal Stresses and Form-Induced Normal Stresses

For a heterogeneous turbulent flow, the SA streamwise and vertical Reynolds normal stresses are expressed as $\langle \sigma_{uu} \rangle = \rho \langle \overline{u'u'} \rangle$, and $\langle \sigma_{ww} \rangle = \rho \langle \overline{w'w'} \rangle$, respectively. Similarly, the streamwise and vertical form-induced normal stresses are $\langle \sigma_{fuu} \rangle = \rho \langle \widetilde{uu} \rangle$ and $\langle \sigma_{fww} \rangle = \rho \langle \widetilde{ww} \rangle$, respectively.

According to Aberle et al. [34], the effects of the spatial heterogeneity in bed roughness on the streamwise time-averaged velocity can be ascertained by analyzing the $\langle \sigma_{fuu} \rangle$ (Figure 10). They observed that, akin to the $\langle \tau_f \rangle$ profile, the $\langle \sigma_{fuu} \rangle$ profile is small away from the crest and gradually increases, as one moves downward toward the crest. Interestingly, they found that for a given bed slope and bed roughness structure, the $\langle \sigma_{fuu} \rangle$ profiles are almost identical for all the discharges. It implies that the shape of the $\langle \sigma_{fuu} \rangle$ profiles are independent of discharge. Further, they compared the $\langle \sigma_{fuu} \rangle$ profiles obtained for different roughness structures, but for a constant bed slope. They argued that the shapes of all the $\langle \sigma_{fuu} \rangle$ profiles are similar, although their absolute magnitudes are different. This suggests that the magnitude of $\langle \sigma_{fuu} \rangle$ profiles is governed by the roughness structure. Then, they analyzed the $\langle \sigma_{fuu} \rangle$ profiles for different bed slopes, keeping the roughness structure identical. The comparison of $\langle \sigma_{fuu} \rangle$ profiles revealed that the variation of bed slope ($S_0 = 0.001$ to 0.01) has a significant impact on the shape of the $\langle \sigma_{fuu} \rangle$ profiles.

The spatial velocity fluctuations $\tilde{u}$ and $\widetilde{w}$ are highly affected by the relative submergence [34]. Hence, to understand the behavioral features of the $\tilde{u}$ with respect to the relative submergence, Koll et al. [28] studied the $\langle \sigma_{fuu} \rangle$ profiles over the original and rotated WGBs. They tested two relative submergences for each bed type: for the original WGB, the relative submergences were taken as 4.4 and 3.2, while, for the rotated WGB, they were 4.5 and 3.3. They noticed that, in the near-bed flow zone, the $\langle \sigma_{fuu} \rangle$ profile increases with an increase in relative submergence. However, away from the bed, the effects of relative submergence diminish. Cooper et al. [36] examined the impact of the relative submergence on both the $\langle \sigma_{fuu} \rangle$ and $\langle \sigma_{fww} \rangle$ profiles in a WGB. In fact, they carried out the analysis for form-induced intensities, $\langle \sigma_{fuu} \rangle^{0.5}$ and $\langle \sigma_{fww} \rangle^{0.5}$. They showed that the SA streamwise form-induced intensity $\langle \sigma_{fuu} \rangle^{0.5}$ profiles exhibit similar shape for all the values of relative submergences. The spatial flow variance is maximum at the middle of the interfacial sublayer, gradually diminishing away from the crest and continuing up to a vertical distance equaling twice the roughness height above the crest.

Further, they observed that between the crest and the vertical distance of twice the roughness height above the crest, the spatial variance is half of its peak value in all the $\langle\sigma_{fuu}\rangle^{0.5}$ profiles, irrespective of the bed roughness. Analysis of the impact of relative submergence on the $\langle\sigma_{fuu}\rangle^{0.5}$ profiles revealed that, for a given vertical distance, the magnitude of $\langle\sigma_{fuu}\rangle^{0.5}$ profile is inversely proportional to the relative submergence. Thus, it confirms that the relative submergence governs the $\langle\sigma_{fuu}\rangle^{0.5}$ profile. By contrast, the results of SA vertical form-induced intensity $\langle\sigma_{fww}\rangle^{0.5}$ profiles inferred that although the shapes of $\langle\sigma_{fww}\rangle^{0.5}$ profiles are similar to those of $\langle\sigma_{fuu}\rangle^{0.5}$ profiles, there is an insignificant difference in the magnitudes of $\langle\sigma_{fuu}\rangle^{0.5}$ profiles owing to the difference in relative submergences. Additionally, they analyzed the $\langle\sigma_{uu}\rangle^{0.5}$ and $\langle\sigma_{ww}\rangle^{0.5}$ profiles for different relative submergences. Akin to $\langle\sigma_{fuu}\rangle^{0.5}$ profiles, the magnitudes of $\langle\sigma_{uu}\rangle^{0.5}$ profiles reduce with an increase in relative submergence, confirming that these profiles are also affected by the relative submergence. Further, they found that the spatial variance in $\langle\sigma_{uu}\rangle^{0.5}$ profiles is approximately half of the spatial variance in the time-averaged streamwise velocity profiles. Moreover, a small variation in $\langle\sigma_{ww}\rangle^{0.5}$ profiles was observed owing to the change in relative submergence. It is important to mention that the spatial variance in $\langle\sigma_{ww}\rangle^{0.5}$ profiles is approximately half of the spatial variance in $\langle\sigma_{uu}\rangle^{0.5}$ profiles and equals the spatial variance in time-averaged vertical velocity profiles. This implies that the spatial flow variance in the streamwise direction is higher than that in the vertical direction.

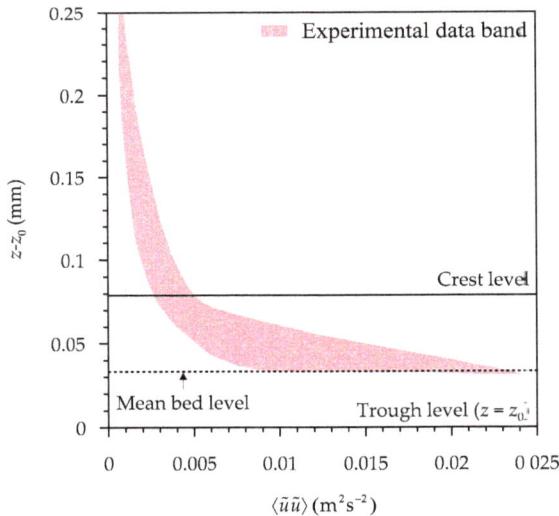

Figure 10. Variations of SA form-induced normal stress $\langle\sigma_{fuu}\rangle$ ($=\langle\widetilde{uu}\rangle/\rho$) with $z-z_0$ in the WGB (data extracted from Aberle et al. [34]).

Considering three types of beds (smooth bed, WGB and SGB), Pu et al. [30] measured the turbulence intensities in streamwise, spanwise and vertical directions. As traditionally found, the turbulence intensities are higher in the rough beds (WGB and SGB) than in the smooth bed. Further, their observations revealed that between the WGB and SGB, the WGB possesses a less even bed roughness structure than that in the SGB. It causes to have larger turbulence intensities and velocity fluctuations in the former than in the latter. It indicates that the WGB can modify the flow turbulence intensity distribution and in turn, the Reynolds normal stresses with respect to the SGB.

Recently, Padhi et al. [32] examined the $\langle\sigma_{uu}\rangle$, $\langle\sigma_{ww}\rangle$, $\langle\sigma_{fuu}\rangle$ and $\langle\sigma_{fww}\rangle$ profiles in a WGB and an SGB. Their analysis showed that owing to the higher WGB roughness height, both u' and w' enhance, resulting in higher values of $\langle\sigma_{uu}\rangle$ and $\langle\sigma_{ww}\rangle$, respectively. Moreover, they also observed that in both the beds, the effects of roughness height are more prominent in the streamwise direction than in the vertical direction. Therefore, the magnitude of $\langle \sigma_{uu} \rangle$ profile, for a given vertical distance, is greater than that of $\langle \sigma_{ww} \rangle$ profile. While comparing the $\langle \sigma_{fuu} \rangle$ and $\langle \sigma_{fww} \rangle$ profiles in both the beds, they found that a higher roughness in the WGB than that in the SGB enhances the $\tilde{u}$ and $\tilde{w}$. As a result, for a given vertical distance, the $\langle \sigma_{fuu} \rangle$ and $\langle \sigma_{fww} \rangle$ profiles in the WGB appear to have higher magnitudes than those in the SGB.

3.4. Effects of Water-Work on Conditional Turbulent Events

Quadrant analysis of two-dimensional velocity fluctuations (u' and w') is usually performed to understand the dynamics of the coherent flow structure in a turbulent flow. In general, in a turbulent boundary-layer flow, the turbulent events generated from the second and the fourth quadrants, termed ejections Q_2 ($-u'$ and $+w'$) and sweeps Q_4 ($+u'$ and $-w'$), respectively, are the dominating events, which govern the turbulence mechanism in the flow. On the other hand, those generated from the first and the third quadrants, termed outward interactions Q_1 ($+u'$ and $+w'$) and inward interactions Q_3 ($-u'$ and $-w'$), respectively, are the weak events, but they can be effective in the context of sediment entrainment [39].

In a WGB, Hardy et al. [27] performed the quadrant analysis to study the relative contribution from each event to the total Reynolds shear stress in governing the turbulent flow. As traditionally observed, their analysis also depicted that the sweeps in the near-bed flow zone are the prevailing events, while the ejections govern in the main flow layer. They however observed more localized flow patterns close to the bed. Near the bed, the ejections and sweeps occur in an alternative manner. In the leeside of a bed undulation, the sweeps govern the flow, while, in the stoss-side of the bed undulation, the ejections are ascendant, as shown in Figure 11. This suggests that the shape of the localized bed topography influences the turbulence characteristics. Furthermore, regarding the outward and inward interactions, they found that the occurrence of these events follows the alternative pattern, as observed for the sweeps and ejections. In the stoss-side of a particle, the outward interactions occur, while in the leeside of a particle, inward interactions prevail. Therefore, as the flow approaches the particle, it decelerates close to the bed; otherwise, it accelerates over or around the particle.

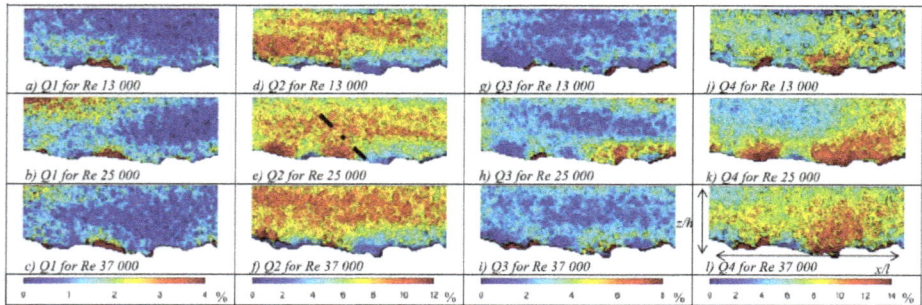

Figure 11. Flow structures as examined with the quadrant analysis: (**a–c**) first quadrant (outward intersections); (**d–f**) second quadrant (ejections); (**g–i**) third quadrant (inward intersections); and (**j–l**) fourth quadrant (sweeps) (data extracted from Hardy et al. [27]).

3.5. Effects of Water-Work on Secondary Currents

Turbulence driven secondary currents of Prandtl's second kind are common in natural streams and arise in the flows in laboratory flumes as well [40], owing to anisotropy in turbulence. In this context, it is pertinent to mention that the secondary currents of Prandtl's second kind are different from those of Prandtl's first kind, which are governed by the streamline curvilinearity induced by the channel boundaries. Presence of secondary currents causes to redistribute the streamflow momentum. The secondary currents are more prominent in a narrow channel, for which the aspect ratio (channel width to flow depth ratio) is less than or equal to 5 [29,40]. On the other hand, in a wide channel having an aspect ratio greater than 5, the flow is deemed to be a two-dimensional with minimal effects of secondary currents at the central portion of flow in a channel [41]. Nevertheless, the secondary currents form cells in both narrow and wide channels.

To study the secondary currents in a WGB, McLelland [42] analyzed the flow characteristics in a WGB for three aspect ratios, such as 5, 10 and 20. They argued that the depth-averaged streamwise velocity could vary about ±5% along the channel cross-section, confirming the existence of secondary currents. The secondary currents forming cells across the channel width have similar dimensions as those along the flow depth and cause the redistribution of mean flow velocity and turbulent kinetic energy. In a WGB, the vectors of secondary currents show the dominance of the vortex with strong counterclockwise motion toward the channel wall-corner and the near-bed current toward the channel centerline. However, this observation in a WGB was fairly opposite to that found in an SGB, where the surface vortex is dominant. Comparing the results of all the aspect ratios, McLelland [42] found that the secondary current cell adjacent to the corner cells for a narrow channel is weaker than those in the intermediate (aspect ratio = 10) and wide channels (aspect ratio = 20). Additionally, in a wide channel, the secondary cells were observed to be stronger in the central portion of the channel. In all the cases, although the flow depth was kept identical for all the aspect ratios, the secondary current cells are of the order of the flow depth, inferring that the bed topography is not solely responsible for the size of secondary current cells. However, it is mainly responsible for the local modifications of the vectors of secondary currents.

4. Conclusions

This article presents the state-of-the-art research on WGBs, highlighting the impact of water-work on the bed topography and the turbulence characteristics. The water-work causes a difference in bed topography between a WGB and an SGB, even for an identical particle-size distribution in both the beds. The orientation as well as the alignment of surface particles in a WGB are adjusted by the flow in such way that the friction angles between the particles are smaller than those in an SGB. Moreover, analysis of the probability distribution function of the bed topographies revealed that the bed roughness structure in a WGB is positively skewed irrespective of sediment feeding rate, discharge and bed mobility conditions. Second- and higher-order structure functions show that a WGB possesses a higher correlation length scale than an SGB, confirming that the former has larger scale bed features than the latter. In addition, the Hurst coefficient in a WGB is higher than that in an SGB, indicating that the roughness structure in the former is more complex than that in the latter.

In the near-bed flow zone, the streamwise velocity in a WGB is more streamlined than that in an SGB owing to the better organized roughness structure in the former than in the latter, wherein they are randomly poised. Additionally, Reynolds and form-induced stresses revealed that the shapes of the turbulence stress profiles are independent of discharge, but dependent on the roughness and relative submergence. Quadrant analysis of turbulent events in WGBs infers that they are governed by the localized bed topography. Further, owing to the presence of the strong secondary currents, the flow cannot be considered as two-dimensional in a WGB, even for a wide channel. Besides, in a WGB, the bottom vortex dominates the flow, while in an SGB, the flow is mainly dominated by the surface vortex.

In essence, owing to the water-work, the roughness structures in WGBs are different from those in SGBs. The order of magnitude of bed roughness as well as turbulence characteristics are also higher in WGBs than in SGBs. Therefore, to obtain accurate results, it is necessary to perform laboratory experiments in WGBs, which resemble natural gravel-bed streams, rather than in SGBs. In addition, the existing results in SGBs are required to be treated carefully, if they are used to predict the resistance to flow, sediment transport, etc. Besides, regarding the scale-effects, in a laboratory experimental study, maintaining the similitude of some of the parameters (such as flow depth, flow velocity, gravel size, roughness, etc.) of the prototype with those of the model is a difficult proposition, as the conditions are often dissimilar. Thus, most experimental models dealing with gravel-bed are analyzed as distorted models, and the question of the applicability of the laboratory experimental results to prototype cases arises. However, Dey [40] and Novak et al. [43] described methods of analyzing the model results of rivers by taking into account the appropriate distortions in the scale ratios of various parameters. Finally, future research should also focus on obtaining corroborating data from prototypes to confirm the scalability of the laboratory results.

Author Contributions: Conceptualization, E.P. and S.D.; formal analysis, E.P.; resources, E.P., S.D. and N.P.; data curation, E.P. and N.P.; writing—original draft preparation, E.P.; writing—review and editing, E.P., S.D., N.P., V.R.D. and R.G.; supervision, S.D., N.P., V.R.D. and R.G.

Funding: This research was partially funded by the JC Bose Fellowship project (JBD).

Acknowledgments: The first author is thankful to the University of Calabria, Italy for the invitation to work in the Laboratorio "Grandi Modelli Idraulici".

Conflicts of Interest: The authors declare no conflict of interest.

References

1. Bathurst, J.C. Theoretical aspects of flow resistance. In *Gravel-Bed Rivers*; Hey, R.D., Bathurst, J.C., Thorne, C.R., Eds.; John Wiley: New York, NY, USA, 1985; pp. 83–108.

2. Bray, D.I. Flow resistance in gravel-bed rivers. In *Gravel-Bed Rivers*; Hey, R.D., Bathurst, J.C., Thorne, C.R., Eds.; John Wiley: New York, NY, USA, 1985; pp. 109–132.

3. Hey, R.D.; Thorne, C.R. Stable channels with mobile gravel beds. *J. Hydraul. Eng.* **1986**, *112*, 671–689. [CrossRef]

4. Furbish, D.J. Conditions for geometric similarity of coarse streambed roughness. *Math. Geol.* **1987**, *9*, 291–307. [CrossRef]

5. Kirchner, J.W.; Dietrich, W.E.; Iseya, F.; Ikeda, H. The variability of critical shear stress friction angle, and grain protrusion in water-worked sediments. *Sedimentology* **1990**, *37*, 647–672. [CrossRef]

6. Robert, A. Boundary roughness in coarse-grained channels. *Prog. Phys. Geogr.* **1990**, *14*, 42–70. [CrossRef]

7. Carling, P.A.; Kelsey, A.; Glaister, M.S. Effect of bed roughness, particle shape and orientation on initial motion criteria. In *Dynamics of Gravel-Bed Rivers*; Billi, P., Hey, R.D., Throne, C.R., Tacconi, P., Eds.; John Wiley: New York, USA, 1992; pp. 23–38.

8. Nikora, V.; Goring, D.; Biggs, B.J.F. On gravel-bed roughness characterization. *Water Resour. Res.* **1998**, *34*, 517–527. [CrossRef]

9. Marion, A.; Tait, S.J.; McEwan, I.K. Analysis of small-scale gravel bed topography during armoring. *Water Resour. Res.* **2003**, *39*, 1334. [CrossRef]

10. Aberle, J.; Nikora, V. Statistical properties of armored gravel bed surfaces. *Water Resour. Res.* **2006**, *42*, W11414. [CrossRef]

11. Buffin-Bélanger, T.; Rice, S.; Reid, I.; Lancaster, J. Spatial heterogeneity of near-bed hydraulics above a patch of river gravel. *Water Resour. Res.* **2006**, *42*, W04413. [CrossRef]

12. Cooper, J.R.; Tait, S.J. Water-worked gravel beds in laboratory flumes: A natural analogue? *Earth Surf. Proc. Land.* **2009**, *34*, 384–397. [CrossRef]

13. Monin, A.S.; Yaglom, A.M. *Statistical Fluid Mechanics: Mechanics of Turbulence*; MIT Press: Cambridge, MA, USA, 1975.

14. Goring, D.; Nikora, V.; McEwan, I.K. Analysis of the texture of gravel beds using 2-D structure functions. In *Proceedings of the IAHR Symposium on River, Coastal, and Estuarine Morphodynamics, Guona, Italy*; Seminara, G., Blondeaux, P., Eds.; Springer: New York, NY, USA, 1999; Volume 2, pp. 111–120.

15. Butler, J.B.; Lane, S.N.; Chandler, J.H. Characterization of the structure of river-bed gravels using two-dimensional fractal analysis. *Math. Geol.* **2001**, *33*, 301–330. [CrossRef]

16. Qin, J.; Ng, S.L. Multifractal characterization of water-worked gravel surfaces. *J. Hydraul. Res.* **2011**, *49*, 345–351. [CrossRef]

17. Nikora, V.; Walsh, J. Water-worked gravel surfaces: High-order structure functions at the particle scale. *Water Resour. Res.* **2004**, *40*, W12601. [CrossRef]

18. Davis, A.; Marshak, A.; Wiscombe, W.; Cahalan, R. Multifractal characterizations of nonstationarity and intermittency in geophysical fields: Observed, retrieved, or simulated. *J. Geophys. Res.* **1994**, *99*, 8055–8072. [CrossRef]

19. Frisch, U. *Turbulence: The Legacy of A. N. Kolmogorov*; Cambridge University Press: New York, NY, USA, 1995.

20. Kolmogorov, A.N. Local structure of turbulence in an incompressible fluid for very large Reynolds numbers. *Dokl. Akad. Nauk SSSR* **1941**, *30*, 299–303. [CrossRef]

21. Alen, J.R.L. *Sedimentary Structures: Their Character and Physical Basis*; Elsevier: New York, NY, USA, 1982.

22. Smart, G.M.; Aberle, J.; Duncan, M.; Walsh, J. Measurement and analysis of alluvial bed roughness. *J. Hydraul. Res.* **2004**, *42*, 227–237. [CrossRef]

23. Bergeron, N.E. Scale-space analysis of stream-bed roughness in coarse gravel-bed streams. *Math. Geol.* **1996**, *28*, 537–561. [CrossRef]

24. Barison, S.; Chegini, A.; Marion, A.; Tait, S.J. Modifications in near bed flow over sediment beds and the implications for grain entrainment. In Proceedings of the 30th IAHR Congress, Thessalonki, Greece, 24–29 August 2003; pp. 509–516.

25. Cooper, J.R.; Tait, S.J. The spatial organisation of time-averaged streamwise velocity and its correlation with the surface topography of water-worked gravel beds. *Acta Geophys.* **2008**, *56*, 614–642. [CrossRef]

26. Legleiter, C.J.; Phelps, T.L.; Wohl, E.E. Geostatistical analysis of the effects of stage and roughness on reach-scale spatial patterns of velocity and turbulence intensity. *Geomorphology* **2007**, *83*, 322–345. [CrossRef]

27. Hardy, R.J.; Best, J.L.; Lane, S.N.; Carbonneau, P.E. Coherent flow structures in a depth-limited flow over a gravel surface: The role of near-bed turbulence and influence of Reynolds number. *J. Geophys. Res.* **2009**, *114*, F01003. [CrossRef]

28. Koll, K.; Cooper, J.R.; Aberle, J.; Tait, S.J.; Marion, A. Investigation into the physical relationship between water-worked gravel bed armours and turbulent in-channel flow patterns. In Proceedings of the HYDRALAB III Joint User Meeting, Hannover, Germany, February 2010.

29. Nezu, I.; Nakagawa, H. *Turbulence in Open-Channel Flows*; CRC Press: Rotterdam, The Netherlands, 1993.

30. Pu, J.H.; Wei, J.; Huang, Y. Velocity distribution and 3d turbulence characteristic analysis for flow over water-worked rough bed. *Water* **2017**, *9*, 668. [CrossRef]

31. Dey, S.; Raikar, R.V. Characteristics of loose rough boundary streams at near-threshold. *J. Hydraul. Eng.* **2007**, *133*, 288–304. [CrossRef]

32. Padhi, E.; Penna, N.; Dey, S.; Gaudio, R. Hydrodynamics of water-worked and screeded gravel beds: A comparative study. *Phys. Fluids* **2018**, *30*, 085105. [CrossRef]

33. Nikora, V.; McLean, S.; Coleman, S.; Pokrajac, D.; McEwan, I.; Campbell, L.; Aberle, J.; Clunie, D.; Koll, K. Double-averaging concept for rough bed open-channel and overland flows: Applications. *J. Hydraul. Eng.* **2007**, *133*, 884–895. [CrossRef]

34. Aberle, J.; Koll, K.; Dittrich, A. Form induced stresses over rough gravel-beds. *Acta Geophys.* **2008**, *56*, 584–600. [CrossRef]

35. Cooper, J.R.; Tait, S.J. Spatially representative velocity measurement over water-worked gravel beds. *Water Resour. Res.* **2010**, *46*, W11559. [CrossRef]

36. Cooper, R.; Aberle, J.; Koll, K.; Tait, S.J. Influence of relative submergence on spatial variance and form-induced stress of gravel-bed flows. *Water Resour. Res.* **2013**, *49*, 5765–5777. [CrossRef]

37. Mignot, E.; Barthelemy, E.; Hurther, D. Double-averaging analysis and local flow characterization of near-bed turbulence in gravel-bed channel flows. *J. Fluid. Mech.* **2009**, *618*, 279–303. [CrossRef]

38. Dey, S.; Das, R. Gravel-bed hydrodynamics: Double-averaging approach. *J. Hydraul. Eng.* **2012**, *138*, 707–725. [CrossRef]

39. Nelson, J.M.; Shreve, R.L.; McLean, S.R.; Drake, T.G. Role of near-bed turbulence structure in bed load transport and bed form mechanics. *Water Resour. Res.* **1995**, *31*, 2071–2086. [CrossRef]

40. Dey, S. *Fluvial Hydrodynamics: Hydrodynamic and Sediment Transport Phenomena*; Springer-Verlag: Berlin, Germany, 2014.

41. Nikora, V.; Roy, A.G. Secondary flows in rivers: Theoretical framework, recent advances, and current challenges. In *Gravel Bed Rivers: Processes, Tools, Environments*; Church, M., Biron, P.M., Roy, A.G., Eds.; John Wiley & Sons Ltd: Chichester, UK, 2012; pp. 3–22.

42. McLelland, S.J. Coherent secondary flows over a water-worked rough bed in a straight channel. In *Coherent Flow Structures at Earth's Surface*; Venditti, J.G., Best, J.L., Church, M., Hardy, R.J., Eds.; John Wiley & Sons Ltd: Chichester, UK, 2013; pp. 275–288.
43. Novak, P.; Cabelka, J. *Models in Hydraulic Engineering*; Pitman: London, UK, 1981.

water

MDPI

Article

Roughness Effect of Submerged Groyne Fields with Varying Length, Groyne Distance, and Groyne Types

Ronald Möws [1],* and Katinka Koll [2]

1   Aller-Ohre-Verband, 38518 Gifhorn, Germany
2   Leichtweiß-Institut für Wasserbau, Technische Universitaet Braunschweig, 38106 Braunschweig, Germany; katinka.koll@tu-bs.de
*   Correspondence: ronald.moews@aller-ohre-verband.de; Tel.: +49-5371-815-415

Received: 16 April 2019; Accepted: 11 June 2019; Published: 14 June 2019

Abstract: Design guidelines were developed for a number of in-stream structures; however, the knowledge about their morphological and hydraulic function is still incomplete. A variant is submerged groynes, which aim to be applicable for bank protection especially in areas with restricted flood water levels due to their shallow height. Laboratory experiments were conducted to investigate the backwater effect and the flow resistance of submerged groyne fields with varying and constant field length and groyne distance. The effect of the shape of a groyne model was investigated using two types of groynes. The validity of different flow types, from "isolated roughness" to "quasi smooth", was analyzed in relation to the roughness density of the groyne fields. The results show a higher backwater effect for simplified groynes made of multiplex plates, compared to groynes made of gravel. The relative increase of the upstream water level was lower at high initial water levels, for short length of the groyne field, and for larger distance between the single groynes. The highest roughness of the groyne fields was found at roughness densities, which indicated wake interference flow. Considering a mobile bed, the flow resistance was reduced significantly.

Keywords: in-stream structures; groyne field; groyne type; backwater effect; flow resistance; friction factor; flow type

---

1. Introduction

Many types of in-stream structures, e.g., stream barbs, bendway weirs, and different kinds of vanes and groynes, were developed aiming to comply with demands for both river training and restoration. An overview of types and related studies is given by Radspinner et al. [1] and, more recently, by Zaid [2]. Studies on ecological benefits e.g., [3], morphological effects e.g., [2,4–10], and flow fields [7,10–15] provide evidence of the structures' potential. Single structures are often used for ecological purposes only, i.e., to increase the variation of river morphology and flow diversity in river restoration projects e.g., [4,10]. For nature-orientated river training, arrangements of groynes are employed, e.g., [2,3,13]. Both single elements and groups of structures require design guidelines for the stability of the structure itself, e.g., [4–6,9], as well as for the specific purpose of implementation. For specific structure types, mainly stream barbs and bendway weirs, design guidelines were developed based on field experience, e.g., [16,17], numerical simulations, e.g., [11], and physical model tests, e.g., [11,13,18]. However, the studies usually conclude that further investigations are required, demonstrating that still some design aspects are not considered.

A feature that cannot be fulfilled by the aforementioned structures is to leave flood water levels almost unaffected. Therefore, submerged groynes were developed for ecologically compatible bank protection especially in areas with restricted flood water levels [2,10,13]. The structure is characterized by a horizontal crest and a height related to the mean low water level at the implementation site. Thus, it is submerged over its full length almost throughout the year and supports the water level especially during low flow conditions. However, during floods the water level rise shall be limited to a minimum. Consequently, knowledge of the backwater height is of high relevance for these structures.

A groyne field can be represented as a large obstacle or a series of small weirs [19]. Yossef [19] considered the field as an obstacle, and developed a formula for calculating the bulk drag coefficient as a function of the blockage ratio and the Froude number, based on experiments with an immobile bed, one groyne setup, and different hydraulic conditions. For applying the formula on a different setup, the flow velocity above the groyne field and in the unblocked part of the cross-section need to be known. The experiments were conducted with nonuniform flow conditions, and thus conclusions cannot be drawn from presented drag force coefficients to assess the development of the backwater with the level of submergence.

Azinfar and Kells [20] investigated the backwater effect of a groyne field, which was simulated by thin plastic plates. They derived a formula relating the effect of the groyne field to the backwater effect of a single groyne. To apply the formula, the flow field caused by a single groyne (backwater height and velocity in that cross-section), its drag force coefficient, and the total relative drag force of the field have to be known. For the latter an empirical formula was provided considering the number of groynes. The effect of a varying groyne field length is implicitly included but not discussed. The authors point out that the formulas are only valid for thin plastic plates and hydraulic conditions within the investigated range (e.g., relative submergence from 1.2 to 2). Transferable findings are that the backwater increases with the number, the spacing of the plates, and increasing submergence, and that the flow resistance of a submerged groyne field is larger than that of a single plate. They concluded that their arrangements are in the wake-interference region following the concept of Morris [21], as the total drag forces were larger than the sum of drag forces caused by an according number of single groynes.

For roughness elements evenly distributed on the bed, Morris [21] distinguished the flow types quasi-smooth (or skimming), wake interference, and isolated roughness flow, which Chow [22] visualized in a sketch. Starting from a dense packing of the roughness elements, the flow resistance of the bed first increases with increasing spacing between the elements until a maximum is reached. For these arrangements, the wakes and vortices at each element interfere with those developed at the neighboring elements, resulting in intense and complex vorticity and turbulent mixing [22]. Further increase of the spacing results in decreasing flow resistance. The spacing can be parameterized by the roughness density c_k, which is the ratio of the upstream projected area of an element to the floor area assigned to that element. According to analysis of numerous studies, the maximum flow resistance, i.e., wake interference flow starts to develop at $c_k = 0.1$ and is maximum at c_k between 0.2 and 0.35, e.g., [23,24]. Comparable results were found using the spatial density of roughness elements [25].

First, recommendations for designing a submerged groyne field for bank protection are given by Mende [13], based on studies in a straight laboratory flume with immobile bed and banks and a simple groyne model made of multiplex (Figure 1a). Möws and Koll [10] improved the groyne model (Figure 1b) and investigated morphological and hydraulic effects on a single submerged groyne in a straight laboratory flume with a mobile bed. In contrast to common results e.g., [5,6,9], the main scour at the groyne head is not attached to the groyne, but located further downstream, which is related to the groyne shape. A similar scouring effect due to the groyne shape was observed by Bressan and Papanicolaou [4] for stream barbs up to a certain level of submergence and by Kadota et al. [8] for permeable submerged groynes. Using the gravel groyne model (Figure 1b) for investigating the effect of geometric groyne parameters [26] and of the position of a submerged groyne [27] on the velocity field in a curved laboratory channel, which finally resulted in recommendations for arranging a submerged groyne field for protecting the outer bank of a bend [2]. However, only one hydraulic boundary condition was adjusted in the latter experiments and thus, the effect of a submerged groyne field on the water level cannot be estimated, yet.

Figure 1. Top view on (**a**) the multiplex groyne model and (**b**) the gravel groyne model.

The aim of the work described in this paper is to investigate the hydraulic roughness of fields of submerged groynes and to assess the effect of model boundary conditions. Using experiments with fixed beds, the effects of roughness density, groyne field length, and the level of simplification of a groyne model on the water level and the friction factor were analyzed. In order to assess the effect of morphological adaptations on the hydraulic roughness, an additional experiment was conducted with mobile bed conditions. Two different groyne models were used as well as varying groyne field lengths and groyne numbers and three levels of submergence. All experiments took place in the hydraulic laboratory of the Leichtweiß-Institute for Hydraulic Engineering and Water Resources.

2. Materials and Methods

The hydraulic roughness of a groyne field, depending on its length and the number of groynes, was investigated in two series of laboratory experiments. The first test series, GF 1, was conducted with a fixed number of groynes (15) and varying distance between the groynes (d_G), resulting in different lengths of the groyne field (l_{GF}). Two types of groynes (Figure 1) were compared: a simplified model of a groyne made of multiplex plates characterized by sharp edges with no sloping from crest to base and a flat top and a second type made from glued gravel (12–16 mm), which resulted in an irregular surface and a slightly permeable body.

In the second series of experiments, GF 2, the length of the groyne field was kept constant (1.5 m). Thus, varying the distance between the groynes resulted in different numbers of groynes. These experiments were conducted only with the gravel groynes.

The fixed bed experiments were conducted in a 30 m long, 60 cm wide, and 70 cm deep hydraulic flume with walls made of glass. On the horizontal flume floor a second floor was built from plywood plates with a length of 15 m and a slope of 0.3%. A single layer of fine gravel (mean diameter $d_m = 3.64$ mm) was glued on the plates to roughen the second floor. A flap gate was installed at the downstream end of the second floor for regulating the water level. The discharge (Q) was controlled by a valve and measured with an inductive discharge meter, and the water depth (h) was measured with a mobile point gauge. The groyne field was located on the left side of the flume always starting at $x = 7$ m (position of groyne toe), with the origin of the longitudinal coordinate $x = 0$ at the upstream end of the second floor. The upstream reach of 7 m length was chosen to ensure fully developed flow conditions. The groynes were installed on top of the rough bed. The groyne height (h_G) of both groyne types was 2.5 cm to investigate relative submergence up to 6. The groyne width (w_G) was 6 cm and the projected length of the groyne (l_P) was 20 cm (1/3 of the flume width). The angle of inclination was chosen to 60° against stream direction (Figure 2) as several studies e.g., [13,18,26] recommend this angle for best protection of a bank.

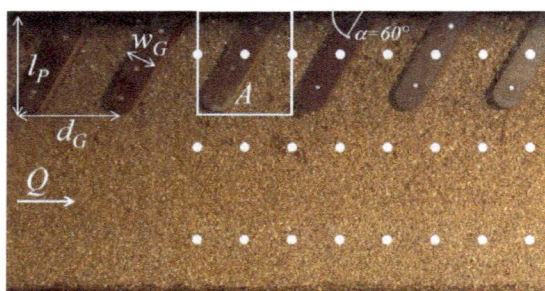

Figure 2. Example of a groyne field setup section with groyne dimension parameters. The white dots exemplarily indicate positions of water level measurements and the square shows the related ground Area (*A*) of a single groyne.

In experiments without groynes, the boundary conditions for uniform flow conditions at water depths h_N = 5, 10, and 15 cm were determined by adjusting the flap gate and the discharge until the target water depth was constant along the flume. The resulting relative submergence for the groyne experiments was 2, 4, and 6. The settings of in total 51 experiments are summarized in Table 1. The roughness density (c_k) is calculated as the ratio of the projected area (A'), which is the product of the groyne height and the projected length, and the related ground area of a groyne (*A*, see Figure 2) (Equation (1)).

$$c_k = \frac{A\prime}{A} = \frac{h_G l_P}{d_G l_P} \tag{1}$$

with m = number of groynes, d_G = groyne distance, d_G/l_P = aspect ratio, l_{GF} = groyne field length, c_k = roghness density, h_N = uniform flow water depth without groynes, Fr = Froude number, and Q = discharge.

Table 1. Experimental parameters.

Series	Groyne Type	m (-)	d_G (cm)	d_G/l_P (-)	l_{GF} (m)	c_k (-)	h_N (cm)	Fr (-)	Q (L/s)
			10	0.5	1.4	0.25			
			12.5	0.625	1.75	0.2			
			15*	0.75	2.1*	0.17*		0.59,	11.6,
GF 1	MultiplexGravel	15	20	1	2.8	0.125	5, 10, 15	0.65,	35.7,
			25	1.25	3.5	0.1		0.66	69.7
			35	1.75	4.9	0.07			
			60**	3	7.8**	0.04**			
		16	10	0.5		0.25			
GF 2	Gravel	11	15	0.75	1.5	0.17	5, 10, 15	0.59,	11.6,
		6	30	1.5		0.083		0.65,	35.7,
		4	50	2.5		0.05		0.66	69.7
GF 3	Gravel	10	20	0.645	1.8	0.125	10	0.59	58.6

* only multiplex groynes, ** only 14 groynes.

In GF 1 the runs with the largest distance between the groynes were carried out with only 14 groynes because of the limited length of the second floor of 15 m.

Water levels were measured in three sections in flow direction: along the center of the groynes (10 cm distance to the left flume wall), along the middle of the flume, and in 10 cm distance to the right flume wall (Figure 2). The water depths were averaged over the cross-section to analyze the bulk effect of a groyne field.

The hydraulic roughness of a groyne field is parameterized by the Darcy–Weisbach friction coefficient f (Equation (2)). For the energy slope, the local energy height was determined by summing up the water depth, averaged across the flume, and the velocity height within each cross-section. The local mean flow velocities were calculated with the continuity equation. A linear trend was fitted to the energy heights starting at the cross-section with maximum water depth and ending at the first cross-section downstream of the groyne field. As the flume walls were much smoother than the flume bed, the hydraulic radius is replaced by the water depth h_{max}.

$$f = \frac{8gh_{max}S_E}{u_m^2} \tag{2}$$

with h_{max} = maximum water depth along the flume, S_E = energy slope, and u_m = mean flow velocity calculated with h_{max}.

The mobile bed experiment (GF 3) was carried out in a 90 cm wide and 60 cm deep tilting flume (slope = 0.3%). A 10-cm-thick layer of fine gravel (mean diameter d_m = 3.64 mm) was placed over a length of 15 m in a 20-m-long flume. A set of ten gravel groynes, with a height of 2.5 cm and l_P = 31 cm (approximately one-third of the flume width), was installed with d_G = 20 cm, 60° angled against flow direction, resulting in c_k = 0.125. The first groyne started at x = 8.35 m, with the origin of x being defined as the beginning of the sediment layer. The water depth at uniform flow conditions (without groynes) was set to 10 cm, which corresponded to incipient motion conditions for the bed material. The experiment ran for 7 hours until only minor changes of the bed topography were observed visually. The resulting bed topography was scanned with a high resolution laser scanner from x = 8 to 10.5 m and across the flume from y = 0.09 to 0.81 m. The resolution of the scan was 0.5 mm in the x-direction and 10 mm in the y-direction. Water levels were measured in three sections in flow direction, relatively related to those in the fixed bed experiments.

3. Results and Discussion

The effect of a groyne field on the water level along the flume was very similar for all experiments, and thus is shown exemplarily for one setup and the three hydraulic conditions in Figure 3. Throughout the experiments the highest increase of the water depth ($\Delta h = h_{max} - h_N$) was located about 10 cm upstream of the groyne field at x = 6.9 m, which corresponded to the position of the first groyne head. Deviations in the exact longitudinal position of the maximum water depth were due to the wavy water surface, which became more predominant with increasing approach velocity, i.e., increasing h_N, and to the fixed raster for water depth measurements with a distance of 10 cm. However, a difference between the water depth at x = 6.9 m and the maximum water depth was observed in only two experiments and was 3 and 4 mm, respectively. Thus, this position is used for further analysis. Towards the downstream end of the groyne field the water level decreased following a linear trend. As expected, the water level downstream of the last groyne was in general lower than h_N. Despite the scatter due to waves, it is obvious that the backwater increased with increasing water depth h_N. However, the rise of the backwater height at position x = 6.9 m decreased with increasing discharge, i.e., the difference of Δh was always larger from h_N = 5 cm to 10 cm than from h_N = 10 cm to 15 cm, indicating a diminishing effect of the groynes with increasing submergence.

The influence of the hydraulic boundary conditions becomes evident by plotting the relative increase of the water depth ($\Delta h/h_N$), measured at x = 6.9 m, which is presented in Figure 4 for the experimental series GF 1 (Table 1). The number of groynes was kept constant, and thus the length of a groyne field grew with the distance between the groynes. Accordingly, the backwater increased with the distance between the groynes; however, approaching a constant backwater height independent of d_G and the groyne field length.

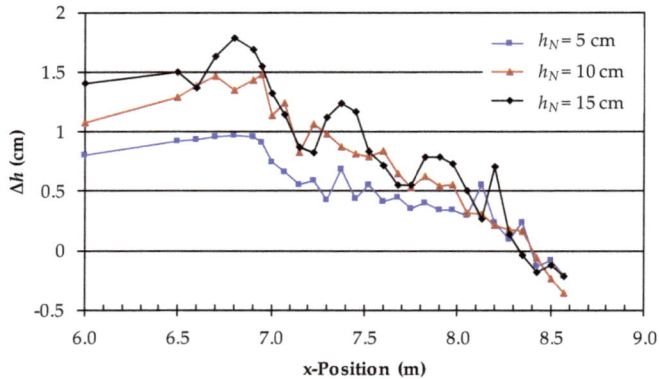

Figure 3. Water level change (Δh) along the main flow direction (groyne field from x = 7 m to 8.5 m) for gravel groynes with distance d_G = 15 cm (GF 2).

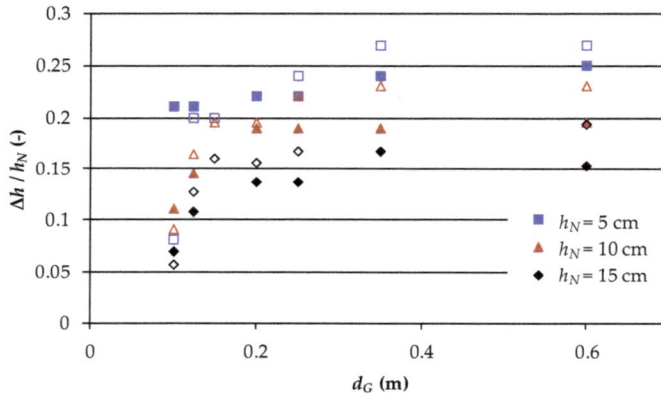

Figure 4. Relative increase of the water depth at x = 6.9 m as a function of the distance between groynes d_G and approach flow in series GF 1 (open symbols = multiplex groynes; filled symbols = gravel groynes).

The change from a steep to an asymptotic increase of $\Delta h/h_N$ occurred at d_G = 15 cm for the experiments with larger submergence. The projected width of a groyne (the distance between head and toe along the x-axis) was 11.55 cm. Thus, the head of a downstream groyne overlapped with or was very close to the toe of the upstream groyne for d_G = 10 and 12.5 cm, resp., which can hinder the development of the overtopping flow, and thus result in lower flow resistance. This effect was observed throughout the experiments except for the gravel groynes with h_N = 5 cm in series GF 1. However, this measured water level may be biased as it should match with the water level measured in the corresponding experiment in series GF 2 (see Figure 5).

Depending on the distance between groynes, the backwater height reached up to 25–27% of the undisturbed water depth for the lowest submergence (h_N = 5 cm) and 15–19% for h_N = 15 cm. The variation is mainly caused by the way of modeling a groyne. The multiplex groynes resulted in higher water depths than the gravel groynes, independent of the level of submergence. The impact was smaller for short distances between the groynes, and the reverse for the shortest groyne distance. The flow field caused by the smooth surface and the sharp edges of the multiplex groynes resulted in higher flow resistances than observed for the irregularly shaped surface and porous body of the gravel groynes. Thus, results from the experiments with strongly simplified substitutes, e.g., the multiplex groynes or plates used by Azinfar and Kells [20] overestimate the groyne influence.

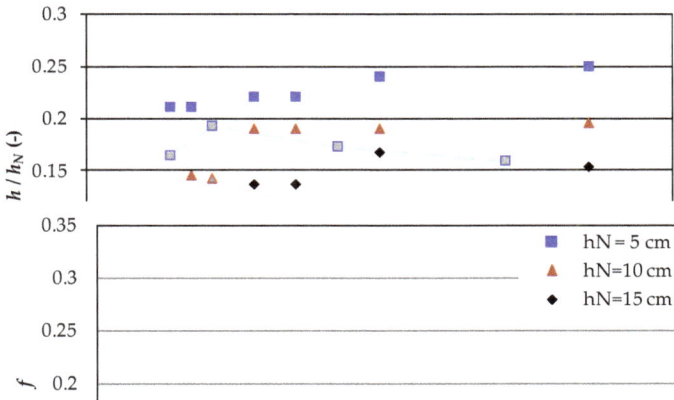

Figure 5. Relative increase of the water depth at x = 6.9 m as a function of the distance between groynes d_G and approach flow for gravel groynes with varying field length (GF 1: filled symbols) and constant field length (GF 2: shaded symbols).

In contrast to the asymptotic increase of the backwater in the experiments with a growing groyne field length (GF 1), the relative increase of the upstream water level reached a maximum, if the groyne field length was kept constant (GF 2) (Figure 5). Independent of the approach flow, the groyne distance d_G = 15 cm resulted in the largest hydraulic roughness. With larger groyne distances the relative backwater decreased asymptotically. The minimum backwater can be expected to correspond to the effect of a single groyne.

Considering a groyne field as a large obstacle as Yossef [19] recommended, its roughness consists of two parameters: the groyne field length and the groyne density within the field. The different relationships, presented in Figure 6, show that the decreasing roughness due to a wider spacing was at least counterbalanced by the increasing roughness of a growing field length. This finding highlights the importance of considering not only the number and spacing of groynes, but of also taking the resulting total length of the groyne field into account. This is important especially for nature-oriented river training as these projects are often planned for river reaches with limited length.

For comparing the experiments with variable and fixed groyne field lengths the friction coefficient f was calculated with Equation (2) (Figure 6). The maximum flow resistance caused by a groyne field was found at d_G = 12.5 and 15 cm, respectively, except for the test run in series GF 1 with gravel groynes and h_N = 5 cm for the aforementioned reason. The multiplex groynes caused larger flow resistance than the gravel groynes. The setup with constant groyne field length (GF 2) resulted in larger friction factors than the setup with variable length (GF 1). However, the friction factors for multiplex and gravel groynes, especially for constant and variable groyne field lengths, followed the same trend. This indicates that the decrease of the energy slope with increasing groyne distance was more distinct for the variable groyne field length than for the constant one. Further investigations require more precise water level measurements to reduce the scatter due to the wavy water surface.

Groyne distances are related to roughness density c_k, which is plotted as second x-axis in Figure 6. The maximum flow resistance corresponded to roughness densities between 0.17 and 0.2 indicating wake interference flow [23,24]. The difference to Canovaro et al. [25], who found maximum flow resistance for spatial densities between 0.2 and 0.4, can be explained with the element height, which is included in c_k, but not in the spatial density.

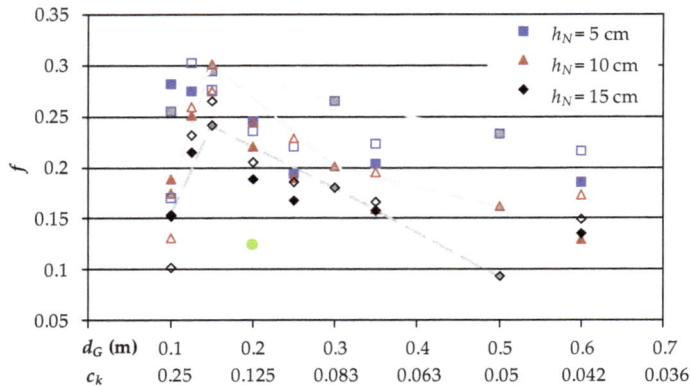

Figure 6. Friction coefficient f of the groyne fields as a function of groyne distance d_G and roughness density c_k, resp. (open symbols = multiplex groynes in series GF 1; filled symbols = gravel groynes in series GF 1; shaded symbols = gravel groynes in series GF 2; green dot: GF 3).

The results show that the concept described by Morris [21] can be applied to the flow over a groyne field, even if the roughness elements are not distributed over the full width of the flume. This finding demonstrates the possibility to use the comprehensive knowledge of flow over rough surfaces to study the hydraulic effects of submerged groyne like structures.

The different backwater heights caused by the simplified multiplex models and the gravel groynes demonstrated that shape matters. For the backwater effect, an overestimation can be considered as positive, as the results would be conservative. However, if simple structures, especially those with sharp edges, are used in morphodynamic studies, misleading conclusions may be drawn. For example, studies of morphological changes due to simplified models report distinct scouring at the head of the groynes, e.g., [6,9]. On the contrary, in mobile bed experiments carried out with more naturally shaped types of single submerged groynes, the main scour was observed downstream of the groyne head [4,8,10] and the stability of the structure was not compromised by a groyne head scour.

The effect of a groyne field made of gravel groynes on the bed morphology and the resulting backwater height was tested for a setup comparable to the run of GF 1 with d_G = 12.5 cm and h_N = 10 cm with respect to the aspect ratio d_G/l_P and the groyne field length (Table 1). Considering the roughness density, the setup was comparable to GF 1 with d_G = 20 cm. The hydraulic conditions led to incipient motion. Sediment transported from upstream towards the groyne field was deposited in the groyne field, while the bed in the unblocked area eroded. The erosive forces became stronger along the groyne field, resulting in scouring at the groyne heads in the middle of the field as well as at the downstream end of the field towards the middle of the flume (Figure 7). The cross-sectionally averaged backwater height at the beginning of the groyne field was Δh = 0.57 cm, while for the two comparable experiments with fixed bed, the water level was increased by 1.45 cm and 1.9 cm, respectively (Figure 4). Although the blockage effect was even larger due to sediment deposition upstream of and within the groyne field, the roughness of the system was compensated distinctly by the areal erosion of the river bed. The resulting water level only increased by approximately 40% and 30%, respectively, compared to the backwater height of the fixed bed experiments. The reduced roughness of the mobile bed system is reflected by the considerably lower friction factor (Figure 6).

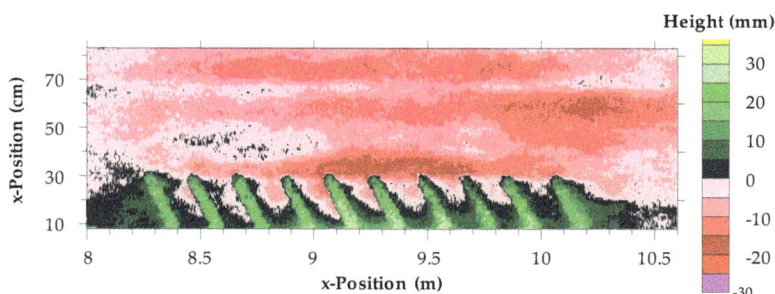

Figure 7. Bed topography at the end of the mobile bed experiment (flow direction from left to right).

4. Conclusions

The hydraulic roughness of submerged groyne fields was investigated for two groyne types: varying groyne distances as well as varying and constant field lengths. For each test geometry, three hydraulic conditions were employed with relative submergences of 2, 4, and 6. The experiments were conducted in a straight flume with a fixed rough bed. One setup was conducted in a flume with a mobile gravel bed at hydraulic conditions for incipient motion to consider the influence of morphological changes on the roughness of the system.

For all setups, the absolute value of the backwater increased with increasing submergence. However, the relative backwater height decreased, confirming the applicability of submerged groynes for situations with restricted flood water levels. The relation between relative water level increase and groyne distance depended on the variability of the groyne field length. With increasing groyne distance and groyne field length (GF 1), the backwater approached a constant height, while for a constant groyne field length (GF 2), the backwater height reached a maximum at a certain groyne distance and decreased with further increasing distance. The different relationships were related to the combined contributions of the groyne density and the groyne field length to the total roughness. Thus, it is important to consider the resulting total length of a groyne field, besides the number and spacing of the groynes, when deriving a formula for calculating the flow resistance of a groyne field. This study investigated only one, quite short groyne field of constant length. Further tests are required to determine for which length the observed effect vanishes.

The change from a steep to an asymptotic increase (GF 1) and the maximum (GF 2), respectively, were found for the groyne distance where the inclined groynes did not overlap anymore. The distance between the head of the next downstream groyne and the toe of the upstream groyne was large enough that the overtopping flow could develop and interact with the flow in the unblocked area, like it is described in [14,15,22]. With increasing distance between the groynes, the losses due to this interaction decreased.

Analysis of the friction factor of a groyne field demonstrated the applicability of the concept of quasi smooth, wake interference and isolated roughness flow, which was developed for uniformly distributed bed roughness, to describe the total roughness of a groyne field. Thus, the comprehensive knowledge of rough bed flow can be used for studying the flow of submerged groyne fields.

These findings were independent of the type of groyne. The simplified shape of the multiplex groyne, especially the sharp edges, resulted in distinctly higher backwater effects compared to the gravel groyne with an irregular surface. It can be assumed, that this effect will affect the results of other studies with substitutes, e.g., plates instead of groyne models. While this can be considered as positive for backwater effects as the results are on the safe side, the results of the morphodynamic studies may lead to misinterpretations. The contribution of the adapted bed topography to the roughness of a river section with groynes resulted in a backwater, which was only approximately one-third of the corresponding backwater measured for fixed bed conditions. The study shows that care must be taken when simplifying the boundary conditions for laboratory experiments, regarding the investigated structure as well as the mobility of the bed, because not only can the magnitude of a measured parameter be affected, but also the functional relationships.

Author Contributions: Conceptualization, R.M. and K.K.; methodology, R.M.; validation, R.M., K.K.; formal analysis, R.M.; investigation, R.M.; resources, K.K.; data curation, R.M.; writing—original draft preparation, R.M.; writing—review and editing, K.K.; visualization, R.M.; supervision, K.K.; project administration, K.K.

Funding: This research received no external funding.

Acknowledgments: Many thanks are given to the students that were involved in the preparation and conduct of the experiments.

Conflicts of Interest: The authors declare no conflict of interest.

References

1. Radspinner, R.R.; Diplas, P.; Lightbody, A.F.; Sotiropoulos, F. River training and ecological enhancement potential using in-stream structures. *J. Hydraul. Eng.* **2010**, *136*, 967–980. [CrossRef]
2. Zaid, B.A. Development of Design Guidelines for Shallow Groynes. Ph.D. Thesis, TU Braunschweig, Braunschweig, Germany, 2017. [CrossRef]
3. Shields, F.D., Jr.; Knight, S.S.; Cooper, C.M. Warmwater stream bank protection and fish habitat: A comparative study. *Environ. Manag.* **2000**, *26*, 317–328. [CrossRef]
4. Bressan, F.; Papanicolaou, T. Scour around a variably submerged barb in a gravel bed stream. In Proceedings of the International Conference on Fluvial Hydraulics, RiverFlow, San José, Costa Rica, 5–7 September; Munoz, R.E.M., Ed.; Taylor & Francis Group: London, UK, 2012; pp. 333–339, ISBN 978-0-415-62129-8.
5. Fang, D.; Sui, J.; Thring, R.W.; Zhang, H. Impacts of dimension and slope of submerged spur dikes on local scour processes-an experimental study. *Int. J. Sediment Res.* **2006**, *21*, 89–100.
6. Hemmati, M.; Daraby, P. Erosion and sedimentation patterns associated with restoration structures of bendway weirs. *J. Hydro-Environ. Res.* **2019**, *22*, 19–28. [CrossRef]
7. Jamieson, E.C.; Rennie, C.D.; Townsend, R.D. 3D flow and sediment dynamics in a laboratory channel bend with and without stream barbs. *J. Hydraul. Eng.* **2013**, *139*, 154–166. [CrossRef]
8. Kadota, A.; Muraoka, H.; Suzuki, K. River-Bed Configuration Formed by a Permeable Groyne of Stone Gabion. In Proceedings of the International Conference on Fluvial Hydraulics, RiverFlow, Çeşme, Turkey, 3–5 September 2008; Altinakar, M.S., Kokpinar, M.A., Aydin, I., Cokgor, S., Kirkgoz, S., Eds.; Kubaba: Ankara, Turkey, 2008; pp. 1531–1540, ISBN 978-605-60136-1-4.
9. Kuhnle, R.A.; Alonso, C.V.; Shields, F.D., Jr. Local scour associated with angled spur dikes. *J. Hydraul. Eng.* **2002**, *128*, 1087–1093. [CrossRef]
10. M#xF6;, R.; Koll, K. Influence of a single submerged groyne on the bed morphology and the flow field. In Proceedings of the International Conference on Fluvial Hydraulics, RiverFlow 2014. Lausanne, Switzerland, 3–5 September 2014; Schleiss, A.J., de Cesare, G., Franca, M.J., Pfister, M., Eds.; Taylor & Francis Group: London, UK, 2014; pp. 1447–1454, ISBN 978-1-138-02674-2.
11. Julien, P.Y.; Duncan, J.R. *Optimal Design Criteria of Bendway Weirs from Numerical Simulations and Physical Model Studies*; Technical Paper; Colorado State University: Fort Collins, CO, USA, 2003.
12. Kuhnle, R.A.; Jia, Y.; Alonso, C.V. Measured and simulated flow near a submerged spur dike. *J. Hydraul. Eng.* **2008**, *134*, 916–924. [CrossRef]
13. Mende, M. Naturnaher Uferschutz mit Lenkbuhnen-Grundlagen, Analytik und Bemessung. Ph.D. Thesis, TU Braunschweig, Braunschweig, Germany, 2014.
14. Sukhodolov, A.N. Hydrodynamics of groyne fields in a straight river reach: Insight from field experiments. *J. Hydraul. Res.* **2014**, *52*, 105–120. [CrossRef]
15. Uijtewaal, W.S.J. Effects of groyne layout on the flow in groyne fields: Laboratory Experiments. *J. Hydraul. Eng.* **2005**, *131*, 782–791. [CrossRef]
16. Lagasse, P.F.; Clopper, P.E.; Pag#xE1;, J.E.; Zevenbergen, L.W.; Arneson, L.A.; Schall, J.D.; Girard, L.G. *Bridge Scour and Stream Instability Countermeasures-Experience, Selection and Design Guidance*, 3rd ed.; HEC-23; US Dept. of Transportation: Fort Collins, CO, USA, 2009; Volume 2.
17. USDA. *Technical Note 23 Design of Stream Barbs-Version 2.0*; Engineering Technical Note No. 23; United States Department of Agriculture, Natural Resources Conservation Service: Portland, OR, USA, 2005.

18. USACE. *Physical Model Test for Bendway Weir Design Criteria*; Report No. ERDC/CHL TR-02-28; U.S. Army Corps of Engineers: Vicksburg, MS, USA, 2002.

19. Yossef, M.F.M. *Morphodynamics of Rivers with Groynes*; Delft University Press: Delft, The Netherlands, 2005; p. 240.

20. Azinfar, H.; Kells, J.A. Drag force and associated backwater effect due to an open channel spur dike field. *J. Hydraul. Res.* **2011**, *49*, 248–256. [CrossRef]

21. Morris, H.M. Flow in rough conduits. *Trans. ASCE* **1955**, *120*, 373–410.

22. Chow, V.T. *OPEN-Channel Hydraulics*; McGraw-Hill Inc.: New York, NY, USA, 1959.

23. Dittrich, A.; Koll, K. Velocity field and resistance of flow over rough surfaces with large and small relative submergence. *Int. J. Sediment Res.* **1997**, *3*, 21–33.

24. Dittrich, A. *Wechselwirkung Morphologie/Strömung Naturnaher Fließgewässer*; Habilitation at Universität Karlsruhe (TH): Karlsruhe, Germany, 1998.

25. Canovaro, F.; Paris, E.; Solari, L. Effects of macro-scale bed roughness geometry on flow resistance. *Water Resour. Res.* **2007**, *43*, 1–17. [CrossRef]

26. Zaid, B.A.; Koll, K. Sensitivity of the flow to the inclination of a single submerged groyne in a curved flume. In *Hydrodynamic and Mass Transport at Freshwater Aquatic Interfaces*; GeoPlanet: Earth and Planetary Sciences; Rowiń, P.M., Marion, A., Eds.; Springer: Heidelberg, Germany, 2016; pp. 245–254. [CrossRef]

27. Zaid, B.A.; Koll, K. Experimental investigation of the location of a submerged groyne for bank protection. In Proceedings of the International Conference on Fluvial Hydraulics, RiverFlow 2016, Saint Louis, MO, USA, 11–14 July 2016; Constantinescu, G., Garcia, M., Hanes, D., Eds.; Taylor & Francis Group: London, UK, 2016; pp. 1286–1292, ISBN 978-1-138-02913-2. [CrossRef]

water

MDPI

Article

Geometric Characteristics of Spur Dike Scour under Clear-Water Scour Conditions

Li Zhang [1,2,*], Pengtao Wang [1,2], Wenhai Yang [1,2], Weiguang Zuo [2], Xinhong Gu [2] and Xiaoxiao Yang [2]

[1] Water Conservancy and Hydropower Engineering, Hohai University, Nanjing 210098, China; wpt@ncwu.edu.cn (P.W.); yangwenhai@ncwu.edu.cn (W.Y.)

[2] Water Institute of Civil Engineers, North China University of Water Resources and Electric Power, Zhengzhou 450046, China; zuoweiguang@ncwu.edu.cn (W.Z.); guxh2018@126.com (X.G.); smileyang123@hotmail.com (X.Y.)

* Correspondence: zhangli1234@ncwu.edu.cn

Received: 19 April 2018; Accepted: 15 May 2018; Published: 24 May 2018

Abstract: Factors such as flow and sediment characteristics affecting the spur dike's local depth of erosion have yielded considerable research results, but there is little discussion of the geometry of the spur dike's local scour holes. This study focuses on the spatial characteristics of the geometry of local scour holes in straight-wall spur dikes. The discussion shows that the spur dike arrangement clearly changes the plane geometry of the scour hole. The maximum scour depth has a power function relationship with the area of the scour hole and the scour hole-volume. Moreover, the ratio of the product of the maximum scour depth and the plane area of the scour hole to the scour hole-volume is a fixed constant. The average slope of upstream of the scour hole and along the axis direction of the spur dike is slightly larger than the angle of repose of the sediment, the slope distribution of the scour holes profiles presents an inverted "U" type, and its profile morphology and slope distribution have geometric similarity. This distribution also reflected that, the interaction between the downward flow and the horseshoe vortex inside the scour hole leads to the formation of a cusp.

Keywords: spur dike; scour; scour depth; scour holes; morphology

1. Introduction

Spur dikes are a typical hydraulic structure building, which have the function of protecting river banks from flow scouring and improving the habitat area of aquatic organisms, and therefore, are widely used in river regulation and water ecological restoration projects, and so on. After the spur dike was arranged in the river, the original flow characteristics and sediment movement of the river section were changed, resulting in scouring of the riverbed around the spur dike. Discussion on the problem of scouring, on the one hand, it is conducive to predicting the geometric characteristics of scour hole, including the degree of vertical dimension and longitudinal dimension erosion, for the optimization of spur dike engineering parameters and design, etc. On the other hand, the physical environmental indicators near spur dikes, such as the deep pool/shoal, and the flow structure, are also the technical bases for improving the assessment of habitat area of aquatic organisms. In recent years, scholars pay more attention to the vertical dimension erosion degree of scour hole in spur dikes, that is, each physical factor affects the scour depth adjustment characteristic, but there is little discussion of the results of the morphology spatial dimension characteristic of scour holes; it is urgent to discuss this problem. The related research results undoubtedly provide technical support for the river regulation and ecological restoration of spur dikes.

Maximum scour depth is one of the important parameters of engineering design. There is a consensus that the characteristics of flow and sediment, the size and type of buildings, etc., affect the law of maximum scour depth [1–3]. In addition to the parameters of the maximum scour depth, the geometric parameters of scour holes in spur dikes also include the plane area of scour hole, scour hole-volume, and plane geometry dimension, etc. Kuhnle et al. [4] pointed out that the ratio of scour hole-volume and the maximum scour depth of a spur dike is approximately constant, due to the impact of flow shallowness and the alignment angle. However, Fael et al. [5] pointed out that the ratio of scour hole-volume and the maximum scour depth increases with the increase of flow shallowness, and further raise an empirical formula for predicting the plane area and volume of scour hole by maximum scour depth. Haltigin et al. [6] predicted the geometric characteristics of scour hole according to the geometry characteristic dimension of scour holes. Williams et al. [7] pointed out the flow shallowness and relative coarseness did not affect the geometric characteristics of the scour hole, and the water resistance rate of the spur dike had an impact on it, however, the geometric characteristics of the scour hole affected by the water resistance rate were not discussed.

The angle of repose of sediment not only reflects the morphology of the scour hole, but also involves the prediction of it and the discussion of the local vortex characteristics [8,9]. One aspect is the discussion of the relationship between the angle of repose of sediment and the slope of the scour hole. For example, Kothyari [10] proposed that the profile slope of a scour hole is equal to the angle of repose of sediment. Zhang [11] pointed out that when the scour reached the equilibrium state, the slope upstream and downstream of a spur dike remains relatively stable, and the slope is approximately the angle of repose of sediment. Karami [12] considered that not only the average slope upstream of the scour hole was equal to the angle of repose of sediment, but also the upstream slope of the scour hole was steeper than the downstream. Some scholars use numerical simulation techniques to predict the depth and shape of local scour holes, based on the feature that the angle of repose of sediment is equal to the slope of the scour hole [8]. However, Zhang [13] pointed out that the angle of repose of sediment is slightly greater than the slope of scour hole. The second is the discussion of the morphological characteristics of the scour holes. Such as Muzzammil [9], who pointed out the existence of a cusp and two distinct slopes in the scour hole. Diab et al. [14] paid attention to the morphological characteristics of each azimuthal of the local scour hole of the bridge pier, and pointed out that the profile of the various scour holes in the pier had similar characteristics. Williams et al. [7] pointed out that the whole process of scour evolution remained similar. Bouratsis et al. [15] discussed, in detail, the average slope of each azimuthal of the bridge pier scour hole, and considered that the average slope distribution characteristics of all azimuths present approximately a sine function. Although there are many discussions, they are discussed in terms of the average slope of the scour hole profile, and the relationship between the slope of scour hole profile and the angle of repose of sediment is still debated, and the profile slope distribution feature is also indistinct.

To sum up, the morphological characteristics of the scour hole are a specific problem in the study of scour mechanisms, which involve different physical factors affecting the erosion characteristics of two dimensional directions of the scour holes, and are still indistinct. In this paper, we choose spur dike as the research object, and discuss this specific problem by carrying out flume experiments and improving the observation method.

2. Experimental Setup and Procedures

The experiment was carried out in a circular flume, whereby the flume is 50 m long and 0.8 m wide. The experimental observation area was located in the middle of the flume, and was 30 m long. The sand adopted uniform sand, and the median particle size d_{50} is 0.2 mm, 0.7 mm, and 1.0 mm respectively; the non-uniform coefficient $\sigma_g = 1.14 \sim 1.3$. The schematic diagram of experimental plane and profile are shown in Figure 1.

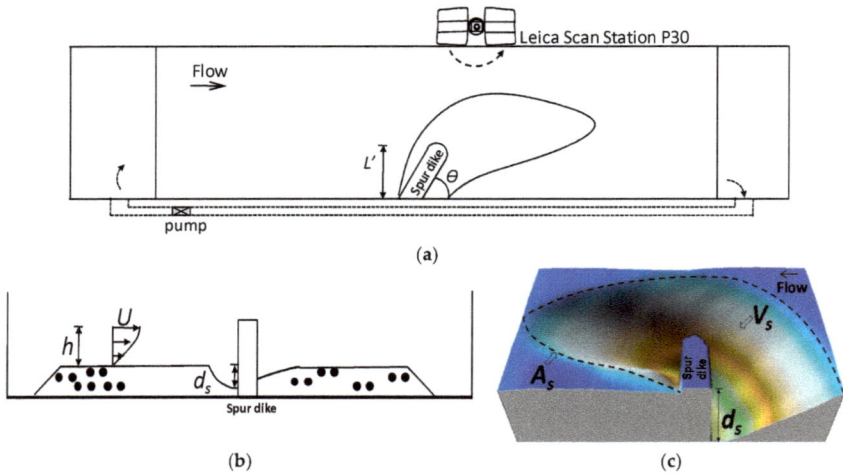

Figure 1. (a) Schematic diagram of flume alignment; (b) Schematic diagram of profiles alignment; (c) Geometric characteristics parameters.

In the practical engineering application, three typical alignment forms are used in the spur dike: $\theta = 90°$, $\theta < 90°$, and $\theta > 90°$; the $\theta > 90°$ alignment, is often applied to submerged spur dikes, while the other two types of alignment are applied to non-submerged spur dikes, such as the Yellow River and the Yangtze River, the spur dikes in $\theta = 90°$ and $\theta < 90°$, as the main alignment. To make the discussion more fitting to the actual engineering application, this study focuses on the non-submerged types, so the design of spur dike angles of alignment were $\theta = 90°$, $\theta = 60°$, and $\theta = 30°$. The length of the spur dikes, L, are respectively 0.12 m and 0.2 m. The structure of spur dikes adopted a vertical wall spur dike with semi-circular type.

Flow depth of spur dike upstream $h = 0.08\sim0.3$ m, with clear-water scouring conditions, and a designed flow intensity $U/U_c = 0.85$, the upstream flow velocity U was measured by an acoustic Doppler velocimeter (ADV); U_c is the incipient velocity of sediment, calculated using the Shamov formula [16]. The design of flow, sediment characteristics, and working conditions are shown in Table 1.

$$U_c = \sqrt{\frac{\rho_s - \rho}{\rho} g d_{50} \left(\frac{h}{d_{50}}\right)} \tag{1}$$

where ρ_s is the density of sediment; ρ is the density of water; g is the acceleration of gravity; d_{50} is the median diameter of sediment; h is the depth of water.

Before the experiment, the bed of the experimental area is kept flat, and the water was slowly stored in the flume to the design depth, adjusting the speed of the axial flow pump, reaching the design flow strength, and carrying out the scour experiment. The local scour of each working condition was 49 h. After the experiment, the water in the sink was slowly vented. After the river bed was dried, the experimental terrain of the local scour hole was collected.

Observation of the area and volume of scour holes in hydraulic structures is more dependent on the improvement of measuring methods; In recent years, measurement techniques, such as LDS (laser distance sensor) [14,17] and high-resolution 3-D monitoring system [18], have been gradually applied to the observation of scour hole morphology. In order to improve the morphology of scour hole of spur dike, a high-speed laser scanner (Leica Scan Station P30) was used to collect the experimental profiles, and its scanning noise accuracy was 0.5 mm @ 50 m; The maximum ranging error within 1 km does not exceed 1.2 mm. Based on the experimental profiles point cloud data, the morphology of the scour holes of spur dikes were refined, as shown in Figure 1c, using Cyclone, Surfer, and other related computation programs to extract the geometric characteristics parameters of scour holes.

The scour depth, the plane area, and scour hole-volume were all calculated using the base surface of the spur dike without the occurrence of the bed surface. The sand waves and the smaller scouring holes downstream were neglected.

3. Results and Discussion

3.1. Scour Depth

The adjustment characteristics of scour depth of spur dike have already got a consensus [1,2,13,19]. Melville [1] thought that for short spur dikes, $d_s/L = 2K_s$; for intermediate spur dikes, $d_s/(Lh)^{0.5} = 2K_sK_\theta$. K_s is the structural shape coefficient, and K_θ is the alignment angle coefficient. However, when $\theta = 90°$, $L = L'$; when $\theta \neq 90°$, $L' = L \times \sin \theta$. The L' is the length of the spur dike projection. In view of this, for the local scour depth of the short spur dikes, the alignment angle coefficient also needs to be considered. According to the results of experimental observations of various working conditions, the adjustment laws of local depth of erosion are shown in Figure 2.

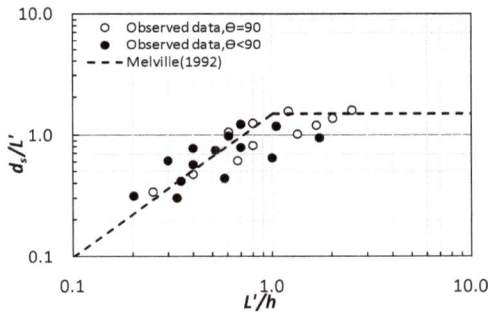

Figure 2. Local scour depth adjustment characteristics.

For short spur dikes, d_s/L' increases as L'/h increases gradually. Regression analysis shows that the slope of the linear relationship is 1.5, close to $2K_s$, and for the straight-wall spur dikes, $K_s = 0.75$. Since the short spur dike only considers the structural shape coefficient, and does not consider the alignment angle coefficient, and when $\theta = 90°$, $K_\theta = 1.0$; $\theta = 60°$, $K_\theta = 0.96$; $\theta = 30°$, $K_\theta = 0.92$. After calculation and correction, it can be seen that the adjustment law is similar to the $\theta = 90°$ arrangement. For the intermediate spur dikes, the upper limit is close to $2K_sK_\theta$ as the L'/h is gradually increased, and is still 1.5.

Based on the above, it is considered that the experimental observations are consistent with the existing research results; in addition, for the short spur dike, the prediction of the local maximum depth of scour and the influence of the alignment angle must also be considered.

3.2. Prediction of Plane Area and Volume of Scour Hole

Based on the principle of dimensionality harmony, the form of $A_s \sim d_s^2$ and $V_s \sim d_s^3$ are usually used to discuss or predict the plane area and scour hole-volume by the maximum scour depth [5,20–22]. In order to facilitate the discussion of this problem, here are the formulas of $A_s \sim d_s^2$; $V_s \sim d_s^3$ is written as the power function relation, that is, $A_s = C_1 d_s^2$; $V_s = C_2 d_s^3$, where C_1, C_2 is the undetermined coefficient, respectively, and its value is related to the influencing factors. For any form of alignment, taking into account the experimental observation results and the existing research results (Tables 1 and 2), C_1, C_2 values and relative coarseness adjustment characteristics can be obtained, as shown in Figure 3.

Table 1. Flow conditions and geometric parameters.

Case	Length L (m)	Alignment Angle θ (°)	Sediment Size d_{50} (mm)	Flow Depth h (mm)	Scour Depth d_s (cm)	Area of Plane A_S (cm²)	Volume V_S (cm³)	L'/d_{50}	L'/h	A_S/d_s^2	V_S/d_s^3
								Non-Dimensional Parameter			
A1	0.12	90	0.2	0.1	17.0	2589.85	12,608.42	600	1.2	8.961	2.566
A2		60		0.1	13.5	2393.81	11,565.47	520	1.04	13.135	4.701
A3		30		0.1	7.0	897.66	1963.36	300	0.6	18.320	5.724
A4		90		0.15	8.5	798.87	2186.84	600	0.8	11.057	3.561
A5		60			7.9	1108.11	3101.92	520	0.69	17.755	6.291
A6		30			2.8	271.91	315.05	300	0.4	34.682	14.352
A7		90		0.3	5.7	301.72	593.44	600	0.4	9.287	3.204
A8		60			4.8	436.11	992.11	520	0.35	18.928	8.971
A9		30			2.1	141.97	100.65	300	0.2	32.193	10.868
A10	0.2	90			12.3	1692.79	6278.40	1000	0.67	11.189	3.374
A11		60			8.4	870.28	3178.40	865	0.57	12.334	5.363
A12		30			3.4	399.91	494.00	500	0.3	34.594	12.569
B1	0.2	90	0.7	0.1	19.5	4337.44	22,029.48	286	2.0	11.407	2.971
B2		60			13.9	2796.7	13,687.77	247	1.73	14.475	5.097
B3		30			7.2	1610.18	4722.7	143	1.0	31.061	12.653
B4	0.12	90		0.15	15.3	2085.74	10,079.63	171	0.8	8.910	2.814
B5		60			14.3	2216.03	10,156.05	149	0.69	10.837	3.473
B6		30			5.2	1011.51	1736.57	86	0.4	37.408	12.350
B7	0.12	90		0.2	12.6	1701.08	6287.63	171	0.6	137.408	3.143
B8		60			12.1	2112.84	6397.93	149	0.52	14.431	3.601
B9		30			9.6	1793.41	5732.78	86	0.3	27.999	6.480
B10	0.05	90			2.6	82.89	103.11	57	0.25	12.262	5.867
S1	0.20	90	1.1	0.08	20.2	6169.60	30,927.71	182	2.50	15.120	3.752
S2	0.20	90		0.12	18.6	2922.98	18,230.19	182	1.67	8.449	2.833
S3	0.20	90		0.15	17.8	3634.43	20,715.28	182	1.33	11.471	3.673

Table 2. Flow conditions and geometric parameters.

Case	Parameter							Non-Dimensional Parameter			
	Length L (m)	Alignment Angle θ (°)	Sediment Size d_{50} (mm)	Flow Depth h (mm)	Scour Depth d_s (cm)	Area of Plane A_S (cm²)	Volume V_S (cm³)	L'/d_{50}	L/h	A_S/d_s^2	V_S/d_s^3
F1	140	90	1.28	6.5	40.7	50,100	776,000	1094	21.5	30.245	11.510
F2	125			6.6	19.9	7770	34,000	977	18.9	19.621	4.314
F3	125			6.9	29.4	25,370	189,000	977	18.1	29.351	7.437
F4	125			6.6	37.2	41,160	595,000	977	18.9	29.743	11.558
F5	109			7.0	16	4730	19,000	852	15.6	18.477	4.639
F6	109			6.9	27.3	22,340	170,000	852	15.8	29.975	8.355
F7	109			6.6	35.9	36,690	480,000	852	16.5	28.468	10.374
F8	94			6.7	33.4	34,480	385,000	734	14.0	30.908	10.333
F9	94			7.0	24.3	15,920	103,000	734	13.4	26.961	7.178
F10	94			7.0	13.1	3050	9000	734	13.4	17.773	4.003
F11	79			6.9	31.2	26,250	284,000	617	11.4	26.966	9.351
F12	79			7.1	23.1	13,550	88,000	617	11.1	25.393	7.139
F13	79			7.1	10.4	3090	6000	617	11.1	28.569	5.334
F14	64			7.0	29.7	23,170	236,000	500	9.1	26.267	9.008
F15	64			7.0	8	850	2000	500	9.1	13.281	3.906
F16	64			7.2	20.8	12,450	69,000	500	8.9	28.777	7.668
K1	30.5	45	0.8	30.2	18.99	/	106,700	381	1.39	/	15.581
K2	30.5	45		18.6	22.41	/	113,800	381	0.86	/	10.112
K3	15.2	45		30.66	16.68	/	67,630	190	2.84	/	14.573
K4	15.2	45		30.7	26.69	/	185,280	190	2.84	/	9.745
K5	15.2	45		18.45	27.98	/	166,720	190	1.71	/	7.611
K6	15.2	45		18.49	17.16	/	55,690	190	1.71	/	11.021
K7	30.5	90		18.56	22.15	/	135,730	381	0.61	/	12.490
K8	30.5	90		30.0	25.68	/	197,600	381	0.98	/	11.668
K9	15.2	90		18.42	15.45	/	99,160	190	1.21	/	26.888
K10	15.2	90		18.63	9.29	/	26,500	190	1.23	/	33.052
K11	15.2	90		30.23	13.04	/	52,470	190	1.99	/	23.663
K12	15.2	90		30.72	16.2	/	109,130	190	2.02	/	25.668
K13	30.5	90		18.28	25.13	/	143,020	381	0.84	/	9.012
K14	15.2	135		18.38	30	/	202,510	190	1.70	/	7.500
K15	15.2	135		18.4	17	/	54,350	190	1.70	/	11.062
K16	15.2	135		30.43	20.9	/	95,530	190	2.82	/	10.464
K17	15.2	135		30.46	28.47	/	260,370	190	2.82	/	11.283

Note: The cases F1–F16 series of data references [5]; The cases K1–K17 series of data references [23]. "/" indicates no observation data.

From the statistical data in Tables 1 and 2, seeing that $L'/d_{50} = 57 \sim 1094$; $h/L' = 0.2 \sim 21.5$. Affected by this, C_1 and C_2 have a slightly larger range of fluctuations, the averages are 20.5 and 8.0 respectively, as shown in Figure 3. With regard to the classification of spur dike type [1], it is considered that this result is also applicable to short and intermediate spur dikes.

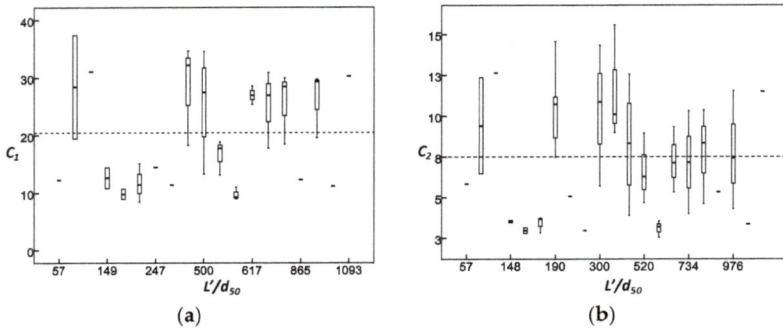

(a) (b)

Figure 3. (a) C_1 adjustment characteristics; (b) C_2 adjustment characteristics.

Regarding scour hole plane area and volume prediction, due to the different impact factors, the research results on the value of the coefficient are not uniform. For example, Kuhnle [23] considered $C_2 = 12.11$; Rodrigue [22] considered $C_2 = 3.87$, and the difference is slightly larger. According to the C_1 and C_2 adjustment characteristics, $C_1 = 20.5$ and $C_2 = 8.0$ are selected, and the plane area and volume of the scour holes are predicted according to the maximum scour depth.

As can be seen from Figure 4, $A_s = 20.5 \, d_s^2$ and $V_s = 8.0 \, d_s^3$ can be used to predict the plane area and volume of scour holes by the maximum scour depth.

The error in predicting the plane area of the scour hole by using the maximum scour depth is slightly larger in contrast; this is because smaller scour holes occur downstream of scour holes when sediment particles are thinner (e.g., $d_{50} = 0.2$ mm). Where, ignored the impact of these small scour holes, which are basically within the range of $\pm 15\%$, which is considered reasonable; see Figure 5.

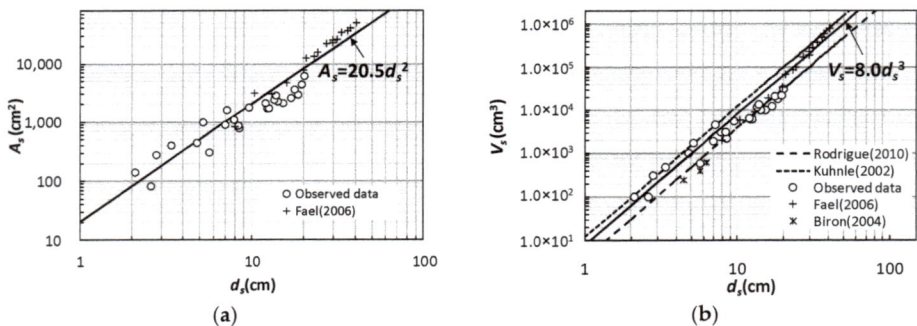

(a) (b)

Figure 4. (a) The relationship between maximum scour depth and plane area of scour hole; (b) The relationship between maximum scour depth and scour hole-volume.

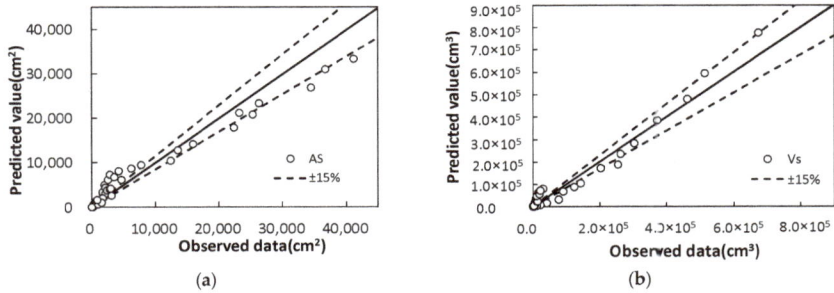

Figure 5. (**a**) The error distribution of scour holes' plane area; (**b**) The error distribution of scour hole-volume.

3.3. The Morphology of Scour Holes

Due to the fact that the alignment of the spur dike changes the erosion of longitudinal dimension scour holes, it is therefore bound to cause changes in the morphology of scour hole. In order to better understand the scour morphology of spur dike, the three-dimensional morphology of the scour hole was reconstructed and visualized according to the experimental point cloud data of each condition. Shown in Figure 6 are selected typical conditions to display.

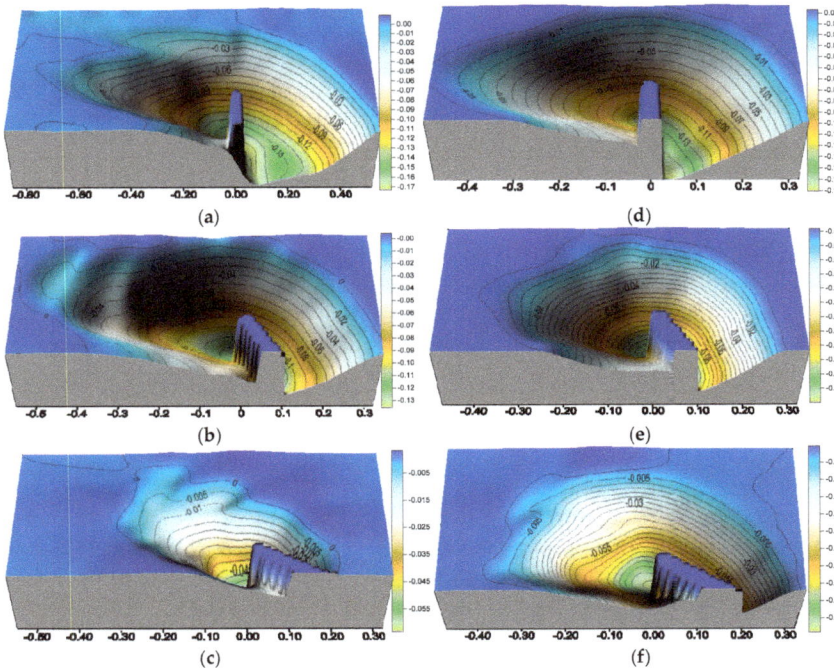

Figure 6. Three-dimensional structure of scour holes in typical conditions (axis unit: m). (**a**) $d_{50} = 1.1$ mm; $L'/d_{50} = 182$; $\theta = 90°$; (**b**) $d_{50} = 0.2$ mm; $L'/d_{50} = 520$; $\theta = 60°$; (**c**) $d_{50} = 0.2$ mm; $L'/d_{50} = 300$; $\theta = 30°$; (**d**) $d_{50} = 0.7$ mm; $L'/d_{50} = 171$; $\theta = 90°$; (**e**) $d_{50} = 0.7$ mm; $L'/d_{50} = 149$; $\theta = 60°$; (**f**) $d_{50} = 0.7$ mm; $L'/d_{50} = 86$; $\theta = 30°$.

For the non-cohesive sand, the morphology of the scour hole in the spur dike is regular and smooth [24]. This phenomenon can also be observed for the scour hole downstream of a rigid bed for non-cohesive sand, while for cohesive materials, the scoured bed forms a non-regular pattern [25]. With the decrease of alignment angle, the position of maximum scour depth obviously changed, and its plane shape gradually changed from oval to triangle. The presence of sand waves downstream from the scour hole was also observed. However, because the bed surface that has not been flushed is used as the reference plane, the sand waves above the reference plane are ignored during the calculation of the volume and other parameters, and the visualization of the local scour hole geometry.

In three-dimensional geometry space, the scour hole-volume is equal to the product of the plane area of the scour hole and the maximum scour depth, that is, $V_s \sim A_s d_s$, but the morphology of the scour hole is not regular in geometry. The ratio between V_s and $A_s d_s$ is still indistinct. Based on the results of experimental observation and previous research, the regulation laws of V_s and $A_s d_s$, under the influence of various factors, are discussed and defined. See Table 1 and Figure 7.

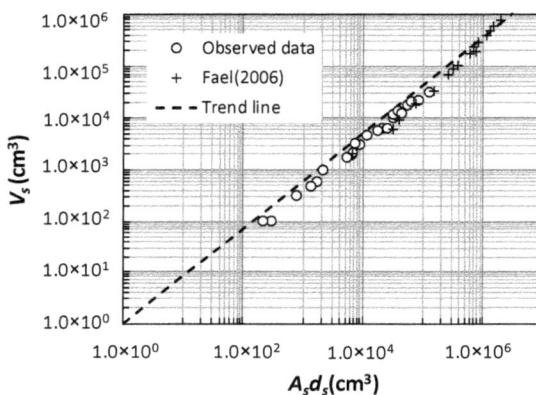

Figure 7. The relationship between the maximum scour depth and the scour hole-volume.

The scour is in equilibrium state, and there is a linear relationship between V_s and $A_s d_s$. Under the influence of those factors, the slope of the linear relationship is the constant of $V_s/A_s d_s$, and regression analysis showed that its value was 0.32; in addition, we easily found that C_1 and C_2 mean ratio is also closer to this constant. Therefore, it can be concluded that the scour hole-volume has a proportional constant with the product of the plane area and the maximum scour depth, and this characteristic also reflects the geometric similarity of the scour hole morphology, which can also be seen from Figure 6.

3.4. The Profile Morphology of the Scour Holes

According to the experimental profile of each case, the scour hole profiles of each azimuthal were extracted. The azimuthal alignment is shown in Figure 8, where, α_i is the azimuth, i = 1, 2, 3; α_1 is defined as scour hole upstream; α_2 is defined as along the spur dike axis direction; and α_3 is defined as the downstream of the scour hole. R_i is the radius of the scour hole corresponding to each azimuthal angle, that is, the width of the scour hole plane.

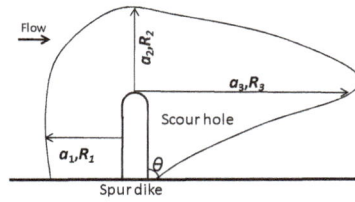

Figure 8. Schematic diagram of each azimuthal plane alignment of the scour holes.

Notice, according to Section 3.1 relative coarseness and scouring depth adjustment law, the two types of alignment $\theta = 90°$ and $\theta < 90°$ were chosen, respectively, and the morphology of the profiles of the scour holes are shown in different directions, as shown in Figure 9.

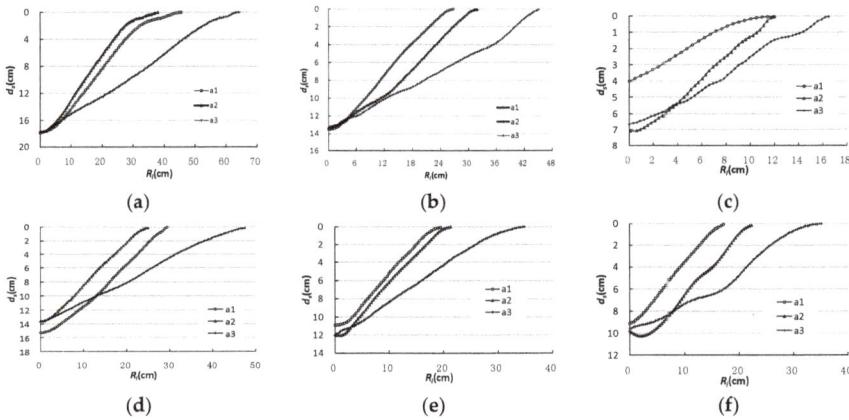

Figure 9. The profile morphological characteristics of each azimuthal of the scour holes. (**a**) $d_{50} = 1.1$ mm; $L'/d_{50} = 182$; $\theta = 90°$; (**b**) $d_{50} = 0.2$ mm; $L'/d_{50} = 520$; $\theta = 60°$; (**c**) $d_{50} = 0.2$ mm; $L'/d_{50} = 300$; $\theta = 30°$; (**d**) $d_{50} = 0.7$ mm; $L'/d_{50} = 171$; $\theta = 90°$; (**e**) $d_{50} = 0.7$ mm; $L'/d_{50} = 149$; $\theta = 60°$; (**f**) $d_{50} = 0.7$ mm; $L'/d_{50} = 86$; $\theta = 30°$.

As can be seen from Figure 9a–f, compared with a1, a2, and a3 three-dimensional profile shape of the scour hole, it can be found that upstream of the scour hole and along the axis direction of the spur dike, the scour hole radius and the profiles are relatively close; downstream of scour hole, the radius is relatively increased, and the slope was significantly slower. Comparing the influence on relative coarseness and the alignment angle of spur dikes, although scour depth and the azimuth of the scour holes radiuses have obvious differences, upstream of scour hole and along the axis direction of spur dike, the profile morphology is similar; the downstream profile of scour holes are also similar. Comparing the scour depth of each azimuth, we see that with the decrease of alignment angle, the position of maximum scour depth gradually transits from the upstream of the spur dike to the head of it.

Diab [14] discussed the different azimuthal profile morphologies of bridge pier scour holes, and pointed out that their profile morphology was geometrically similar. Williams [7] pointed out that the relative coarseness did not affect the geometric similarity of pier scour holes profile morphology.

Results of Tafarojnoruz [26] also show that some kinds of pier scour countermeasures, e.g., pier slot, may not significantly change the similarity of the scour shape.

Comparing the scour hole profile morphology characteristics of the spur dikes, all of them have geometric similarity, which shows that the relative coarseness and spur dike alignment do not affect the profile morphology characteristics of the hole, and they have geometric similarity.

3.5. The Profiles Slope of the Scour Holes

According to the profile morphology characteristics of each azimuthal scour hole, in accordance with the trigonometric relationship, the slope (in terms of angle, the same as below) of any point of the scour hole profiles is calculated piece by piece, according to the trigonometric function, that is, $\varphi_{ij} = \arctan(\Delta ds / \Delta R_{ij})$, where, Δds is the vertical height difference between adjacent points of scour hole profiles; ΔR_{ij} is the horizontal distance between two adjacent points; φ_{ij} is the slope value of any point on the slope of the scour holes, $i = 1, 2, 3$, respectively, corresponding to α_1, α_2, α_3; and j is the number of calculation, $j = 1, 2, 3, ..., n$.

The sediment median particle size is unchanged, and the angle of repose is constant. The sediment angle of repose, which references [13], where $d_{50} = 0.2$ mm, $\varphi = 33.1°$; $d_{50} = 0.7$ mm, $\varphi = 34.8°$; $d_{50} = 1.1$ mm, $\varphi = 35.4°$; and where φ is the sediment angle of repose.

The dimensionless parameter φ_{ij} / φ undoubtedly reflects the difference between the profile slopes and the angle of repose of sediment; the dimensionless parameter R_{ij} / R_i is normalized to the radius of each scour hole. Therefore, the dimensionless parameter R_{ij} / R_i and φ_{ij} / φ relationship pare reflect the azimuth of the profile slopes distribution characteristics. The profile slope distribution of each azimuth is shown in Figure 10.

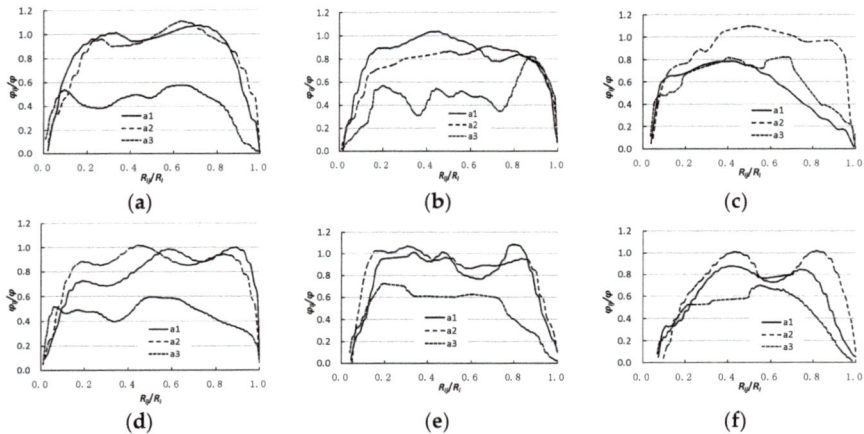

Figure 10. The profile slope distribution of each azimuth of the scour holes. (**a**) $d_{50} = 1.1$ mm; $L'/d_{50} = 182$; $\theta = 90°$; (**b**) $d_{50} = 0.2$ mm; $L'/d_{50} = 520$; $\theta = 60°$; (**c**) $d_{50} = 0.2$ mm; $L'/d_{50} = 300$; $\theta = 30°$; (**d**) $d_{50} = 0.7$ mm; $L'/d_{50} = 171$; $\theta = 90°$; (**e**) $d_{50} = 0.7$ mm; $L'/d_{50} = 149$; $\theta = 60°$; (**f**) $d_{50} = 0.7$ mm; $L'/d_{50} = 86$; $\theta = 30°$.

Comparing the profile slopes at each azimuthal scour hole, it is easily found that regarding the scour hole upstream and along the axis direction of spur dike, the scour hole profile steepness is closer; in the scour hole downstream, profile slope morphology is similar, but the slope has obviously slowed down. Comparing the impact of relative coarseness and the alignment angle, the relationship between R_{ij} / R_i and φ_{ij} / φ shows that although the slope distribution is slightly different, but presents from small to large, and then the trend is reduced, showing an inverted "U" distribution. This shows that profile slopes distribution of scour hole also has geometric similarity.

Both for the bridge piers and the spur dikes, the existing research results indicated that the average slope upstream of the scour hole is approximately equal to the sediment repose angle, and the profile slopes upstream are greater than those downstream [12,27]. The φ_{ij} / φ ratio indicates that the slope of a certain distance in the scour hole is approximately equal to the sediment repose angle, and the average value is smaller than the sediment repose angle. Zhang [13] pointed out that the slope ratio of upstream and downstream angles of the scour hole is constant, about 0.5. The experimental results further indicate that the ratio of the average slope upstream and along the axis direction of spur dike, and the average slope downstream of the scour hole, with a mean of 0.6, and the discussion results, are relatively close.

Unger and Hager [28] pointed that the interaction between the downward flow and the horseshoe vortex inside the scour hole leads to the formation of a cusp, separating the region in the scour hole mainly shaped by the downward-flow and the region shaped by the horseshoe vortex and the separation vortexes. Indeed, there is a cusp in the slope of the scour hole, which leads to a sudden change in the slope of the scour hole. The overall shape is an inverted "U" pattern. The scour mechanism depends fundamentally on the downward flow, and not on the intensity of the horseshoe vortex, as argued by Shen et al. [29]. However, considering the slope distribution characteristics of the scour hole, it is closely related to the distribution pattern of the horseshoe vortex system, and the eddy current size and intensity.

4. Conclusions

Based on the flume experiment, the effect of relative coarseness and the alignment of spur dikes on the morphological characteristics of scour hole are discussed. The results show the following:

Under clear-water scour conditions, for the vertical wall spur dike with semi-circular type. Using $C_1 = 20.5$ and $C_2 = 8.0$, it is reasonable to predict the plane area and volume of the scour hole by maximum scour depth. There is a fixed proportional relationship between the product of the plane area, and the maximum scour depth and the scour hole-volume, and the constant is 0.32, which has geometric similarity. With the decrease of alignment angle, the position of maximum scour depth gradually approached the head of spur dike. The arrangement of the spur dike significantly changed the position of the local maximum scour depth and the plane shape of the scour hole. With the decrease of alignment angle, there is a gradual transition from an approximate ellipse to an approximately triangular shape. The position of maximum scour depth gradually approached the head of spur dikes.

Affected by the relative coarseness and the alignment of the spur dikes, the average slope upstream of the scour hole and along the axis direction of the spur dike is slightly larger than the angle of repose of the sediment, and both are steeper than the downstream angle of the scour hole; the ratio of the average slope of the upstream and along the axis direction of spur dike, and the average slope downstream of the scour hole, ranged from 0.5 to 0.86. The slope distribution of the scour hole profiles present an inverted "U" type distribution, and the profile morphology and slope distribution have geometric similarity.

Author Contributions: Li Zhang, Pengtao Wang conceived the experiments; Pengtao Wang and Wenhai Yang designed and performed the experiments; XinhongGu and XiaoxiaoYang recorded experimental data; Weiguang Zuo analyzed the data and drew pictures; Li Zhang analysis and the interpretation of the results, wrote the paper.

Acknowledgments: This work was funded by the National Natural Science Foundation of China (51579103); Funding for the open project fund of Key Laboratory of the Yellow River sediment of the Ministry of Water Resources (201805).

Conflicts of Interest: The authors declare no conflicts of interest.

References

1. Melville, B.W. Local scour at bridge abutments. *J. Hydraul. Eng.* **1992**, *118*, 615–631. [CrossRef]
2. Melville, B.W. Bridge abutment scour in compound channels. *J. Hydraul. Eng.* **1995**, *121*, 863–868. [CrossRef]
3. Lee, S.O.; Sturm, T.W. Effect of sediment size on physical modeling of bridge pier scour. *J. Hydraul. Eng.* **2009**, *135*, 793–802. [CrossRef]
4. Kuhnle, R.A.; Alonso, C.V.; Shields, F.D., Jr. Local scour associated with angled spur dikes. *J. Hydraul. Eng.* **2002**, *128*, 1087–1093. [CrossRef]
5. Fael, C.M.S.; Simarro-Grande, G.; Martín-Vide, J.P.; Cardoso, A.H. Local scour at vertical-wall abutments under clear-water flow conditions. *Water Resour. Res.* **2006**, *42*. [CrossRef]

6. Haltigin, T.W.; Biron, P.M.; Lapointe, M.F. Predicting equilibrium scour-hole geometry near angled stream deflectors using a three-dimensional numerical flow model. *J. Hydraul. Eng.* **2007**, *133*, 983–988. [CrossRef]

7. Williams, P.D. Scale Effects on Design Estimation of Scour Depths at Piers. Ph.D. Thesis, The University of Windsor, Windsor, ON, Canada, 2014.

8. Bihs, H.; Olsen, N.R.B. Numerical modeling of abutment scour with the focus on the incipient motion on sloping beds. *J. Hydraul. Eng.* **2011**, *137*, 1287–1292. [CrossRef]

9. Muzzammil, M.; Gangadhariah, T. The mean characteristics of horseshoe vortex at a cylindrical pier. *J. Hydraul. Res.* **2003**, *41*, 285–297. [CrossRef]

10. Kothyari, U.; Garde, R.; Ranga Raju, K. Temporal variation of scour around circular bridge piers. *J. Hydraul. Eng.* **1992**, *118*, 1091–1106. [CrossRef]

11. Zhang, H.; Nakagawa, H.; Kawaike, K.; Baba, Y. Experiment and simulation of turbulent flow in local scour around a spur dyke. *Int. J. Sediment Res.* **2009**, *24*, 33–45. [CrossRef]

12. Karami, H.; Ardeshir, A.; Saneie, M.; Salamatian, S.A. Prediction of time variation of scour depth around spur dikes using neural networks. *J. Hydroinform.* **2012**, *14*, 180–191. [CrossRef]

13. Zhang, L.; Sun, K.Z.; Xu, D.P. Morphological evolution of spur dike local scour hole and the scour balance critical condition. *J. Hydraul. Eng.* **2017**, *48*, 545–550.

14. Diab, R.M.A.E.A. *Experimental Investigation on Scouring around Piers of Different Shape and Alignment in Gravel*; TU Darmstadt: Darmstadt, Germany, 2011.

15. Bouratsis, P.; Diplas, P.; Dancey, C.L.; Apsilidis, N. Quantitative Spatio-Temporal Characterization of Scour at the Base of a Cylinder. *Water* **2017**, *9*, 227. [CrossRef]

16. Wang, H.; Tang, H.W.; Xiao, J.F.; Wang, Y.; Jiang, S. Clear-water local scouring around three piers in a tandem arrangement. *Sci. China Technol. Sci.* **2016**, *59*, 888–896. [CrossRef]

17. Oscar, L.P.; fleger, F.; Zanke, U. Characteristics of developing scour-holes at a sand-embedded cylinder. *Int. J. Sediment Res.* **2008**, *23*, 258–266.

18. Bouratsis, P.P.; Diplas, P.; Dancey, C.L.; Apsilidis, N. High-resolution 3-D monitoring of evolving sediment beds. *Water Resour. Res.* **2013**, *49*, 977–992. [CrossRef]

19. Coleman, S.E.; Lauchlan, C.S.; Melville, B.W. Clear-water scour development at bridge abutments. *J. Hydraul. Res.* **2003**, *41*, 521–531. [CrossRef]

20. Cheng, N.S.; Chiew, Y.M.; Chen, X. Scaling an alysis of Pier-Scouring Processes. *J. Eng. Mech.* **2016**. [CrossRef]

21. Biron, P.M.; Robson, C.; Lapointe, M.F.; Gaskin, S.J. Deflector designs for fish habitat restoration. *Environ. Manag.* **2004**, *33*, 25–35. [CrossRef] [PubMed]

22. Rodrigue-Gervais, K.; Biron, P.M.; Lapointe, M.F. Temporal development of scour holes around submerged stream deflectors. *J. Hydraul. Eng.* **2010**, *137*, 781–785. [CrossRef]

23. Kuhnle, R.A.; Alonso, C.V.; Shields, F.D. Geometry of scour holes associated with 90 spur dikes. *J. Hydraul. Eng.* **1999**, *125*, 972–978. [CrossRef]

24. Debnath, K.; Chaudhuri, S. Effect of suspended sediment concentration on local scour around cylinder for clay-sand mixed sediment beds. *Eng. Geol.* **2011**, *117*, 236–245. [CrossRef]

25. Dodaro, G.; Tafarojnoruz, A.; Sciortino, G.; Adduce, C. Modified Einstein sediment transport method to simulate the local Scour evolution downstream of a rigid bed. *J. Hydraul. Eng.* **2016**, *142*, 04016041. [CrossRef]

26. Tafarojnoruz, A.; Gaudio, R.; Calomino, F. Evaluation of flow-altering countermeasures against bridge pier scour. *J. Hydraul. Eng.* **2012**, *138*, 297–305. [CrossRef]

27. Barbhuiya, A.K.; Dey, S. Local scour at abutments: A review. *Sadhana* **2004**, *29*, 449–476. [CrossRef]

28. Unger, J.; Hager, W.H. Down-flow and horseshoe vortex characteristics of sediment embedded bridge piers. *Exp. Fluids* **2007**, *42*, 1–19. [CrossRef]

29. Shen, H.W.; Schneider, V.R.; Karaki, S. Local scour around bridge piers. *J. Hydraul. Div. Proc. Am. Soc. Civ. Eng.* **1969**, *95*, 1919–1940.

![water logo] *water*

MDPI

Article

The 3-D Morphology Evolution of Spur Dike Scour under Clear-Water Scour Conditions

Li Zhang [1,2,*], Hao Wang [2], Xianqi Zhang [1], Bo Wang [1] and Jian Chen [1]

[1] Water Institute of Civil Engineers, North China University of Water Resources and Electric Power, Zhengzhou 450046, China; zhangxianqi@ncwu.edu.cn (X.Z.); wangbosky99@153.com (B.W.); chenjian@ncwu.edu.cn (J.C.)

[2] Water Conservancy and Hydropower Engineering, Hohai University, Nanjing 210098, China; wanghaohhu@hhu.edu.cn

* Correspondence: zhangli1234@ncwu.edu.cn

Received: 2 September 2018; Accepted: 15 October 2018; Published: 5 November 2018

Abstract: By changing the alignment angle of spur dike, this study focused on the evolution of scour hole morphology in three alignments under clear-water scour conditions, including the 3-D structure of the scour hole, 2-D profile morphological evolution process and the evolution characteristics of the local bed shear stress. The results show that the plane area and volume of the scour hole both are power functions over time, which is similar to the evolution characteristics of scour depth. Local scour includes three stages: The initial stage, development stage and balance stage. The local bed shear stress presents successively: $\tau_b > \tau_c$, $\tau_b = \tau_c$ and $\tau_b < \tau_c$. Based on this characteristic, the evolution mechanism between scour hole morphology and the local bed shear stress is further clarified. Furthermore, although the alignment of the spur dike significantly affects the longitudinal and vertical dimension erosion rates of the scour hole, the scour hole morphology is not only relatively constant but also has a specific proportion, and the evolution process is orderly in the whole process of evolution. To the scouring equilibrium state, the length of the scour hole on the upstream and downstream of the spur dike is approximately in line with the golden section feature. The related results provide technical support for scour parameter design and scour hole protection of spur dike in flood period.

Keywords: spur dike; scour; scour holes; morphology; local bed shear stress

---

1. Introduction

Spur dike is a widely used hydraulic structure in river control engineering. After the spur dike was built, the pattern of the flow is locally changed, for example, the generation, separation and attenuation of surrounding vortices make the flow present strong three-dimensional turbulent characteristics, and the flow structure is very complex. The study of local scour near the spur dike not only has important value for hydraulics, but also has practical guiding significance for the practical engineering application of the spur dike.

Under clean-water scour conditions, the scholars pay more attention to the vertical dimension erosion characteristics of the scour hole, namely the evolution characteristics of scour depth [1,2]. Affected by large and small-time scales, the exponential function relation can be used to describe scour depth evolution characteristics [3,4]. In addition, the power function relation [5] and logarithmic function relation are often discussed [6]. In comparison, the discussion results of scour hole's morphological parameter are slightly smaller. We only reported the influence of flow depth and the alignment angle on the evolution rule of scour hole morphology, and thought that the ratio of scour hole-volume to scour depth was approximately constant over time [7,8]. Oscar et al. [9] pointed out that one of the reasons for the lack of relevant research results is that the observation of scour hole area and volume is more dependent on the improvement of measurement techniques. With the application of new measuring tools, for example, the literature [6] not only reported the evolution process of bridge pier scour hole 3-D structure and visualization, but also proposed the relationship between scour depth and scour hole-volume to present a cubic polynomial function. The scour hole-volume presents the growth trend of the power function over time, which is similar to that of scour depth. Bouratsis et al. [10] also reported the 3-D structure evolution process of bridge piers scour hole and discussed in detail the evolution trend of scour hole-volume with time. Some other studies also present the role of the bridge scour countermeasures on 3D structure of the scour hole pattern [11]. Although it showed the increase of the power function, there were clearly two different growth rates. However, there is no literature on the 3-D structure evolution of the scour hole in the spur dike.

The profile morphological characteristics of the scour hole, including the profile morphological evolution process of the scour hole over time, and spatial dimension characteristics under the scour equilibrium state, are also an area of interest for many scholars [6,12]. With the passing of time, the profile morphology of the scour hole remains constant. Zhang et al. [13] further reported the slope distribution characteristics of the scour hole profile in the spur dike, and thought that its distribution characteristics were closely related to local flow. Although the above literature [6,12,13] has paid attention to the morphological characteristics of scour hole profiles from the time dimension or the spatial dimension, the relevant discussion on profile area of the scour hole with time evolution has not been mentioned. It is important that this parameter is one of the important parameters which cannot be ignored in the local bed shear stress evolution model, namely, the evolution characteristics of the scour hole profile area determines the evolution trend of local bed shear stress [14–17]. At present, more of the literature has reported that the local bed shear stress gradually decreases over time [18–20], the scour reaches the equilibrium state, and its value is close to the constant. The constant is about $(0.3\sim0.5)\ \tau_c$, in which τ_c is the bed critical shear stress [21,22]. However, other scholars pointed out that although the local bed shear stress was close to constant in the state of scouring equilibrium, the evolution process did not completely follow this rule, and the evolution characteristics were not further discussed in detail. It can be seen that based on the morphology evolution of scour hole profile, the evolution trend of local bed shear stress is an urgent issue for discussion.

This study intends to adopt clear-water scouring condition and sets different alignment angles to conduct experimental research, and carry out discussions on the evolution characteristics of scour hole morphology and the regulation of local bed shear stress. Relevant research results will enrich and improve the research contents of the local scour mechanism.

2. Experimental Setup and Procedures

The experiment was carried out in a circular recirculating flume, which is 50 m long and 0.8 m wide. The observation area was located in the middle of the flume and was 30 m long. The design alignment angles of spur dike are respectively $\theta = 150°$, $\theta = 90°$ and $\theta = 30°$. The length of the spur dike is $L = 0.12$ m, which changes with the alignment angle, and the projection length of the spur dike is $L' = L \times \sin\theta$. The structure of spur dike is adopted Vertical wall Spur dike with semi-circular type. For the sketch of experimental plane alignment, see Figure 1a. The sand adopted is uniform sand, with the median particle size $d_{50} = 0.7$ mm, and the non-uniform coefficient $\sigma_g = 1.2$. The flow depth and velocity at the upstream of the spur dike are all constant, at h = 0.15 m; $U = 0.21$ m/s, respectively. The flow intensity $U/U_c = 0.87$, U_c is the incipient velocity of sediment [13], which belongs to Clear-Water scour conditions.

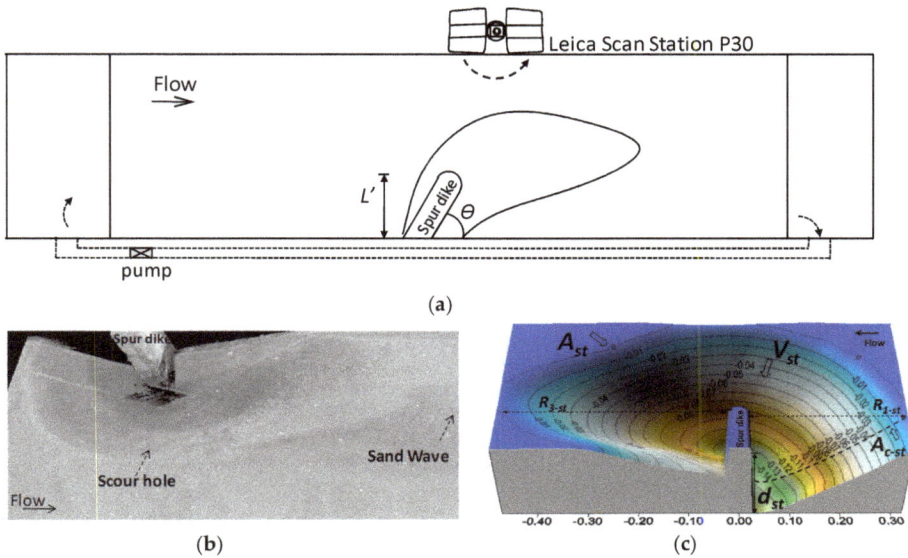

Figure 1. (a) Schematic diagram of flume experiment; (b) the point cloud data of scour hole topography; (c) Geometric characteristic parameters and characteristic length of scour hole (axis unit: m).

Before the experiment, the bed of the experimental area is kept flat, and the water was slowly stored in the flume to the design depth, adjusting the speed of the axial flow pump, reaching the design flow strength, and carrying out the scour experiment. The scouring duration of the experiment was $t = 0.5, 1, 2, 3, 5, 12$ and $T = 48$ h, respectively. After the designed scouring period is completed, stop the experiment, scour hole topography will be collected, and then the topography will be restored to the leveling state for the next scouring duration experiment. In particular, the flume is a closed circulation system, and the designed flow depth and velocity in each case are constant and consistent. Meanwhile, the erosion duration in each case is strictly controlled. These measures ensure the accuracy requirements of the experiment in each case.

The scour hole topography was collected using a high-speed laser scanner (Leica Scan Station P30-High-Definition 3D Laser Scanning Solution), and the scanning noise accuracy was 0.5 mm (within 50 m). The range accuracy is no more than 1.2 mm (within 1 km). The topography of scour experiment is constituted by point cloud data. Leica Cyclone 3D Point Cloud Processing Software, Golden Software Surfer and other related calculation programs are used to reconstruct the three-dimensional morphology of the scour hole and calculate the morphology parameters of the scour hole, such as the area and volume of scour hole. It is noted that due to flow intensity $U/U_c < 1$, the upstream bed of spur dike remains the original bed (zero plane) all the time, and larger sand waves are formed in the downstream of spur dike under all cases, as shown in Figure 1b. In order to facilitate the discussion of the problem, when the three-dimensional morphology of scour hole is visualized, the large-scale sand waves in the downstream are ignored. When calculating the area and volume of the scour hole, zero plane is used as the reference surface. See Figure 1c for details.

3. Experimental Results

3.1. The 3-D Structure of Scour Hole

In order to more intuitively understand the 3-D structure evolution process of scour hole, based on the experimental topographic point cloud data, the 3-D morphology of scour hole at each moment is reconstructed and visualized in three alignments of spur dike, that is $\theta = 150°$, $\theta = 90°$ and $\theta = 30°$ respectively, as shown in Figure 2.

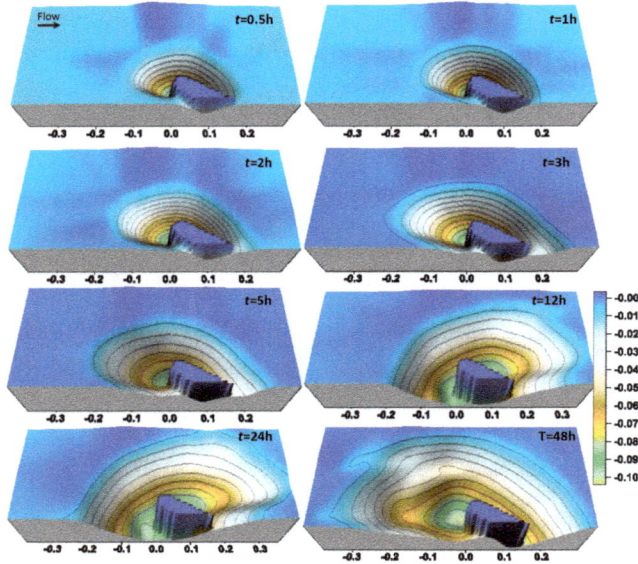

(a)

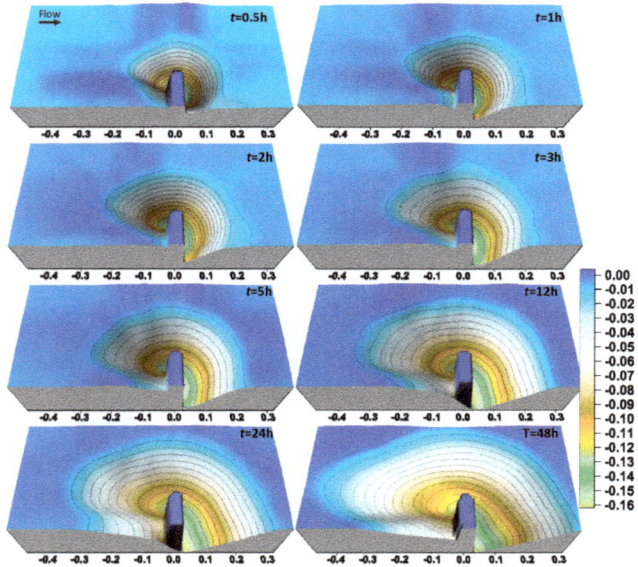

(b)

Figure 2. *Cont.*

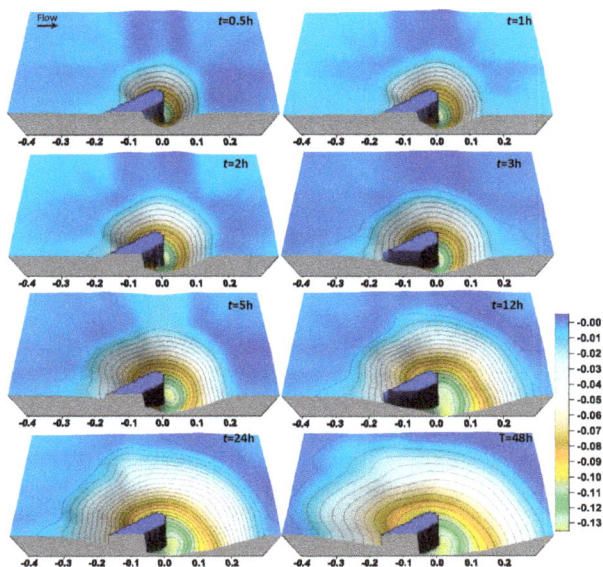

Figure 2. 3-D structure of scour holes (axis unit: m); (**a**) $\theta = 150°$ (**b**) $\theta = 90°$ (**c**) $\theta = 30°$.

It can be seen that both scour depth and the geometric size of the scour hole increase gradually over time. When the balance state was reached, the maximum depth was 13.9 cm, 16.3 cm and 10.7 cm respectively. By contrast, when $\theta = 90°$, both the scour depth and geometry dimensioning of the scour hole are the largest. The scour range of the upstream and downstream of the spur dike is obviously changed due to the different the alignment angle of the spur dike; when $\theta = 150°$, the scour area of upstream is significantly greater than that of downstream; when $\theta = 30°$, the scour area of the spur dike downstream is slightly larger.

To the equilibrium state, when $\theta = 90°$, the 3-D structure of the scour hole is obviously regular, and the plane shape of scour hole is nearly semicircular; For $\theta = 150°$, $\theta = 30°$ (it referred to as $\theta \neq 90°$ alignment, similarly hereinafter), the three-dimensional structure of scour hole is slightly irregular, and the plane shape of scour hole is closer to triangle. Along the axis of the spur dike, or Karman vortex street area, there is a step -in scour hole, at $\theta = 30°$ this alignment is especially clear. This phenomenon should be closely related to the local flow characteristics, as shown in Figure 2a,c; however, there is no literature that discusses this phenomenon.

3.2. The Evolution of Plane Area and Volume of Scour Hole

The flow conditions in each case were calculated, and scour depth, the plane area and volume of scour hole were calculated, as shown in Table 1.

Under clean-water scour conditions, the research results of scour depth characteristics over time are extremely rich [23], this will not be described again. The discussion results of the plane area $A_{st}\sim A_{se}$ and scour hole-volume $V_{st}\sim V_{se}$ over time are slightly deficient. According to the observation results of scour hole geometry parameters during the whole evolution process, the evolution characteristics are discussed.

Table 1. Flow conditions and experimental results.

Alignment Angle θ (°)	Flow Depth h (m)	Projected Length L' (m)	Scour Time (h)	Scour Depth d_{st} (cm)	Area of Plane A_{st} (cm²)	Volume V_{st} (cm³)
			0.5	4.3	196.27	387.76
			1	5.2	287.63	630.80
			2	6.2	357.68	842.34
			3	6.8	431.52	1229.00
150	0.15	0.06	5	7.7	571.99	1836.62
			12	9.6	804.50	3203.94
			24	13.1	1009.68	4674.46
			48	13.9	1285.27	7097.38
			0.5	5.6	234.28	485.68
			1	6.6	417.23	959.99
			2	7.8	605.10	1570.12
			3	8.9	778.07	2266.63
90	0.15	0.12	5	9.8	898.55	2747.63
			12	12.2	1317.20	5002.13
			24	13.7	1698.56	7720.24
			48	16.3	2285.44	12,415.15
			0.5	4.2	219.27	323.66
			1	5.1	290.14	474.57
			2	5.9	396.68	663.13
			3	6.5	464.60	1025.67
30	0.15	0.06	5	7.4	580.11	1522.65
			12	8.6	737.43	2259.77
			24	9.6	917.72	3013.09
			48	10.7	1087.89	4014.74

Similar to the scour depth evolution rule, the area and volume of scour hole also show a power function relation over time, and its expression is as follows:

$$\frac{A_{st}}{A_{se}} = C_1 \left(\frac{t}{T}\right)^m ; \ \frac{V_{st}}{V_{se}} = C_2 \left(\frac{t}{T}\right)^k \tag{1}$$

where, C_1 and C_2 are constants; m and k are indices. Regression analysis showed that, for the three alignments: $\theta = 150°$, $\theta = 90°$, $\theta = 30°$, C_1 and C_2 were close to constant 1; the m values were 0.33, 0.42 and 0.28 respectively; k values were 0.58, 0.68 and 0.45 respectively. Kuhnle et al. [7,8] thought that when describing the evolution characteristics of scour hole-volume, $k = 0.579$–0.653, the results were relatively close. It is believed that it is reasonable for $m = 0.28$–0.42 and $k = 0.45$–0.68 to describe the evolution characteristics of scour hole plane area and volume when $\theta = 30$–$150°$. By comparison, for $\theta = 90°$, the evolution curve is relatively straight, and the values of index m and k are slightly larger. For $\theta \neq 90°$, the evolution curve is relatively curved, and the values of index m and k are slightly smaller, as shown in Figure 3.

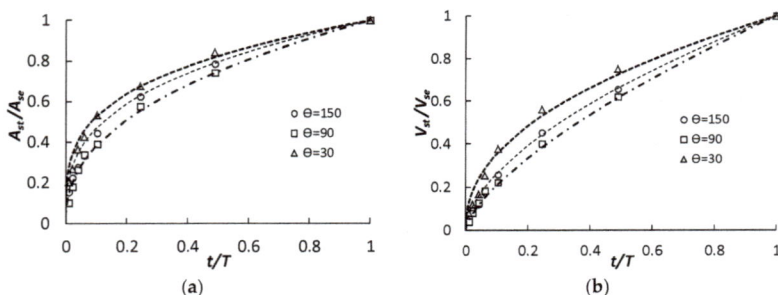

Figure 3. The evolution characteristics (**a**) the plane area of scour hole (**b**) scour hole-volume.

3.3. Difference of the Vertical and Longitudinal Dimension Evolution of Scour Hole

Here, Ast~dstp; Vst~dstq is adopted to discuss the differential characteristics of the vertical and longitudinal dimension evolution rates of scour hole, which was rewritten as the equal sign:

$$A_{st} = C_3 d_{st}{}^p; \ V_{st} = C_4 d_{st}{}^q \tag{2}$$

where, C_3 and C_4 are constants, p and q are exponents. For further deformation of Equation (2), the constant is normalized, so that $A'_{st} = \frac{1}{C_3} A_{st}; \ V'_{st} = \frac{1}{C_4} V_{st}$, the difference discriminant of the vertical and longitudinal dimension evolution rates of scour hole can be obtained:

$$A'_{st} = d_{st}{}^p; \ V'_{st} = d_{st}{}^q \tag{3}$$

where, A'_{st}, V'_{st} are respectively the plane area and volume of the scour hole after normalization. According to the experimental observation results and the values of exponential p and q, the difference between vertical and longitudinal dimension evolution of scour hole is discussed, as shown in Figure 4.

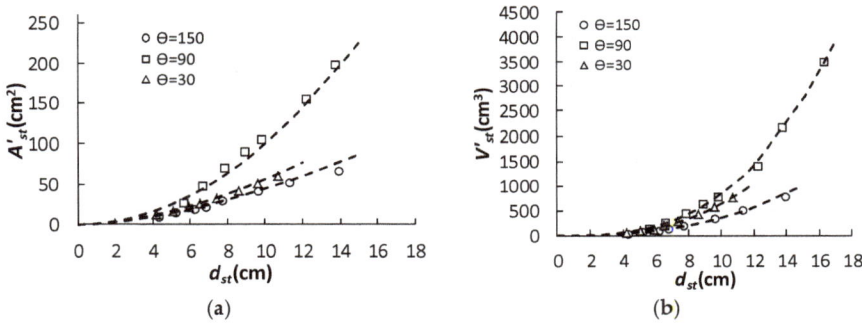

Figure 4. The difference of scour hole geometrical evolution velocity (**a**) the plane area of scour hole (**b**) scour hole-volume.

Regression analysis showed that when $\theta = 150°$, $\theta = 90°$, $\theta = 30°$, the p values were respectively 1.65, 2.0, and 1.75; while the q values were 2.55, 2.92 and 2.8 respectively. If the longitudinal dimension evolution rate of scour hole is greater than that of vertical dimension evolution rate, then $p < 2$, $q < 3$; otherwise, $p > 2$, $q > 3$ [4]. Therefore, it is easy to understand that for $\theta = 90°$, the p and q values were respectively close to 2.0, 3.0; the longitudinal and vertical dimension evolution rates of the scour hole are close to the same. For $\theta \neq 90°$, $p < 2$, $q < 3$; the longitudinal dimension evolution rate of the scour hole is greater than that of vertical dimension evolution rate. It shows that there are also significant differences in the longitudinal and vertical dimension erosion rates of the scour hole in different spur dike alignments, which undoubtedly provides support for the theoretical model of scour evolution proposed by the literature [4].

3.4. Constant Ratio of 3-D Morphological

Zhang et al. [13] discussed the three-dimensional morphological spatial dimension characteristics of scour hole by conducting flume experiments and thought that the scour hole-volume has a proportional constant with the product of the plane area and the maximum scour depth, which is 0.32, and thought that the 3-D morphology of scour hole has a certain geometric similarity. However, it is still unclear whether the ratio relationship between V_{st} and $d_{st}A_{st}$ still has the same characteristics throughout the evolution process of local scour. The discussion was conducted according to the experimental observation results (in Table 1), as shown in Figure 5.

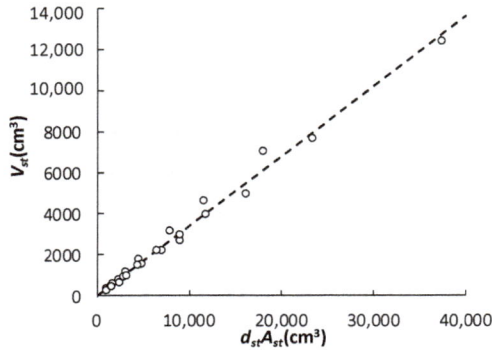

Figure 5. Ratio between V_{st} and $d_{st}A_{st}$.

It can be seen that V_s and $A_s d_s$ show a linear relationship with a slope of 0.34, that is, the ratio of V_s and $A_s d_s$ is still close to the proportional constant, which is the same as the existing observation results. Therefore, it can be considered that the morphology still has a certain degree of geometric similarity in the whole evolution process of the scour.

3.5. The Morphology and Area of Scour Hole Profile

According to the experimental topography of each evolution stage (Figure 2), the profile morphology of scour hole on the upstream and downstream of the spur dike was extracted, and its orientation alignment and the calculation sketch of scour hole profile area $A_{c\text{-}st}$ were shown in Figure 1c for details. Figure 6 shows the morphology adjustment situation of scour hole profile on upstream and downstream of spur dike during the whole evolution process.

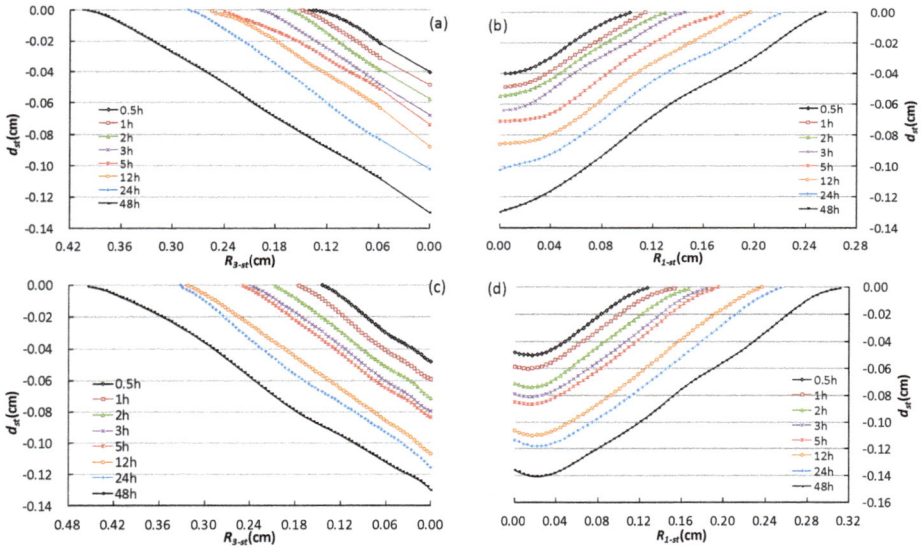

Figure 6. *Cont.*

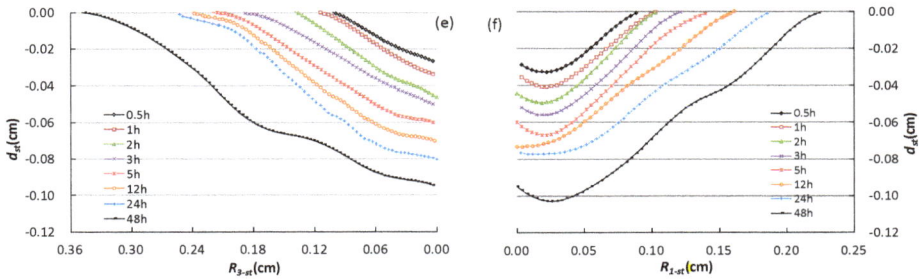

Figure 6. Characteristics of scour hole profile morphology over time (**a,b**) $\theta = 150°$, (**c,d**) $\theta = 90°$, (**e,f**) $\theta = 30°$.

The discussion in the above section shows that there is a slight difference between the upstream and downstream scour range of different spur dikes. Geometrically, the upstream and downstream profile morphology of the scour hole is close to the inverted triangle with R_{1-st} and R_{3-st} as the base edge and scour depth d_{st} as the height. If it satisfies the following equation, it is in line with the golden section feature:

$$\frac{R_{3-st}}{R_{1-st}} = \frac{R_{3-st} + R_{1-st}}{R_{3-st}} \tag{4}$$

where the plane geometric lengths of scour hole on the upstream and downstream of spur dike are respectively represented by R_{1-st} and R_{3-st}.

The results of the experimental show that, when $\theta = 150°$, the mean values of Equation (4) both ends are 1.43 and 1.70 respectively; When $\theta = 90°$, 1.48–1.67; $\theta = 30°$, 1.35–1.74. All are close to the constant 1.618. That is to say, in the entire evolution process, the downstream and upstream geometric characteristic length of the spur dike approximately conforms to the golden section feature, and it is the closer to the scour balance state, the more obvious this characteristic is. The scour hole plane morphology of the bridge pier also accords with a similar characteristic [24].

Geometrically, the upstream and downstream profile morphology of the scour hole is close to an inverted triangle. From the perspective of time dimension, its evolution process always remains constant. In fact, the distribution of horseshoe vortex core and vortex system in the scour hole of spur dike upstream is stable [10,25]. Therefore, the slope of profile remains relatively constant throughout the evolution over time. Equation (4) further indicates that the scour hole profile not only remains relatively constant throughout the evolution process, but also has a certain proportion relation, which forms a stable state when it is close to the golden section feature.

According to the profile morphology of scour hole in each evolution stage, the profile area of scour hole at each evolution stage is calculated (see Table 2), and the relation between scour depth and the profile area of the scour hole is drawn, as shown in Figure 7.

The regression analysis shows that the relation between scour depth d_{st} and the profile area of scour hole A_{c-st} approximately conforms to the power function of index 2. Throughout the evolution process, the ratio of A_{c-st}/d_{st}^2 decreases, but both are greater than or equal to 1; when the scour reached a balance state, the dimensionless parameters A_{c-st}/d_{st}^2 were 1.07, 1.21 and 1.01 respectively for $\theta = 150°$, $\theta = 90°$, $\theta = 30°$, as shown in Table 2.

Table 2. Scour depth and the profile area of scour hole.

Alignment Angle θ (°)	Scour Time t (h)	Scour Hole d_{st} (cm)	Profile Area $A_{c\text{-}st}$ (cm^2)	Initial Area A_0 (cm^2)	Shear Tress τ_b (N/m^2)	Dimensionless Parameter $A_{c\text{-}st}/d_{st}^2$
	0.5	4.03	21.98		0.429	1.35
	1	4.90	30.97		0.360	1.29
	2	5.49	37.96		0.323	1.26
150	3	6.42	46.95	2.918	0.289	1.14
	5	7.11	56.94		0.260	1.13
	12	8.55	80.92		0.215	1.11
	24	10.26	118.88		0.174	1.13
	48	12.99	180.82		0.137	1.07
	0.5	5.04	36.96		1.456	1.46
	1	6.03	49.95		0.588	1.38
	2	7.39	68.93		0.511	1.26
90	3	8.10	84.92	9.48	0.436	1.30
	5	8.65	90.91		0.393	1.21
	12	11.01	146.85		0.379	1.21
	24	11.80	171.83		0.295	1.23
	48	14.07	238.76		0.271	1.21
	0.5	3.25	16.56		0.494	1.57
	1	4.08	23.92		0.411	1.44
	2	4.94	29.44		0.370	1.21
30	3	5.60	36.80	2.918	0.329	1.17
	5	6.69	49.00		0.292	1.09
	12	7.35	60.72		0.251	1.13
	24	7.73	69.00		0.234	1.15
	48	10.31	107.64		0.183	1.01

Note: Local depth values in Tables 1 and 2 are slightly different due to different profile locations.

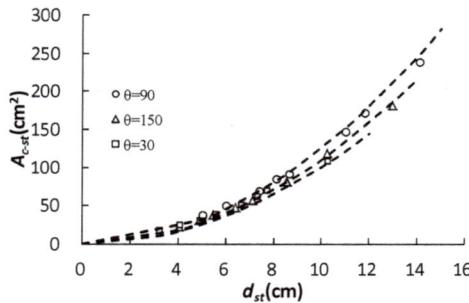

Figure 7. Relation curve of depth and profile area in scour hole.

3.6. The Local Bed Shear Stress Evolution

With time evolution, the scour range increases, the size of horseshoe vortex increases, the intensity of vortex decreases, and the local bed shear stress decreases gradually. Kothyari [14] first proposed the local bed shear stress evolution model, as shown in the following formula:

$$\tau_b = 4\tau_0 \left(\frac{A_0}{A_0 + A_{c-st}}\right)^{0.57} \tag{5}$$

where, A_0 is the initial area of the main vortex at $t = 0$, $A_0 = (\pi/4)(0.28h(L'/h)^{0.85})^2$; A_{c-st} is the area of the main vortex at T, That is the area of a particular profile at upstream of the scour hole; $A_{c-st} = \lambda d_{st}^2\cos\phi$, in which, ϕ is the sediment angle of repose; λ is constant. It can be seen from the ratio of the dimensionless parameter A_{c-st}/d_{st}^2 in the upper section that the process of evolution is $\lambda \geq 1$. However, results of previous studies were all adopted for $\lambda < 1$ [15]. Even consider the difference of sediment particle size, namely assuming $\phi = 30$–$44°$ [6], $\cos\phi < 1$ also exists. Obviously, the λ value should be several times larger. There are reasons to think that $\lambda < 1$ is also necessary to discuss. $d_{50} = 0.7$ mm, the critical friction velocity $u_{*cr} = 0.019$ m/s was calculated, and the upstream flow shear stress was further calculated: $\tau_0 = 0.364$ N/m^2, please refer to the literature [26] for detailed process.

Barbhuiya et al. [22] reported the evolution of local scour from $t = 0$ to T. and the characteristics of local bed shear stress evolution, namely:

$$\tau_b = \beta\tau_c|d_{st} \to d_{se} \tag{6}$$

in which, β is the constant, τ_c is the critical bed shear stress, calculated according to the shields parameter, $\tau_c = 0.34$ N/m^2. In fact, the local erosion evolution process β is not constant, but it is still unclear.

According to Equation (5), in the whole evolution process, $\beta = f(A_{c\text{-}st} \to A_{c\text{-}se})$. The results of this study established and corrected the functional relationship between d_{st}^2 and $A_{c\text{-}st}$. Therefore, further modification of Equation (6) can be used to obtain the relationship describing the evolution characteristics of local bed shear stress:

$$\tau_b = f(d_{st} \to d_{se})\tau_c \tag{7}$$

According to Equation (7), the relationship between τ_b/τ_c and d_{st}/d_{se} was established, and then the characteristics of local bed shear stress evolution at each stage of scour evolution were discussed (see Figure 8).

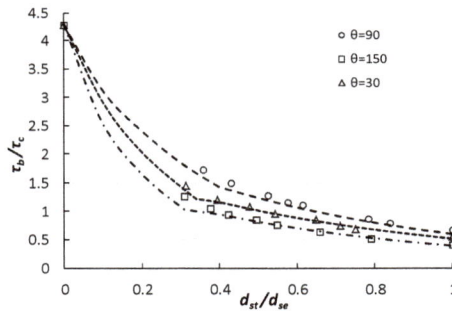

Figure 8. The trend of local bed shear stress evolution.

In Figure 8, clearly shows the adjustment process of the ratio of τ_b/τ_c. With the evolution of scour depth, the overall trend of τ_b/τ_c is decreasing, which approximately conforms to the exponential function.

Referring to three typical stages of local scour evolution [27,28]: The initial stage of the local scour, the ratio of τ_b/τ_c is 4.5 from the upper limit, close to the results discussed in the literature [14], and rapidly decline to 1.0, which indicates that the local bed shear stress decreases rapidly. Dodaro et al. [29] also demonstrates a rapid decrease of bed shear stress with temporal evolution of the scour hole during the initial phase of scour development. When the local scour develops to the equilibrium stage, the ratio of τ_b/τ_c is close to 1.0. At the equilibrium state of the scour hole, the ratio of τ_b/τ_c is close to 0.3–0.6. The results show that the rate of shear stress decrease is slow. The value is consistent with the discussion results in the literature [21,22].

4. Discussion

4.1. Scour Hole Morphological Evolution Rate

Under clean-water scour conditions, it is a clear problem to discuss the morphology of the scour hole from a spatial dimension or a time dimension. Due to the improvement of means of measurement, 3-D structure, 2-D profile morphological evolution and visualization of scour hole have become a reality. Over time, scour depth, the plane area and volume of the scour hole all presented a power function increase (see Table 1), which was confirmed by previous research results [8,10].

It is not difficult to find that when the evolution duration is the same, the percentage of $\theta \neq 90$ for this alignment is slightly larger than that of $\theta = 90°$ in completing the total scour charge (see Figure 6); or the time needed for the former to reach the flush equilibrium state is less than the latter. In short, the morphology evolution of the scour hole is relatively slower for $\theta = 90°$, when $\theta \neq 90°$, the morphology evolution of the scour hole is relatively faster.

Further discussion shows that when $\theta = 90°$, the entire process of evolution, longitudinal and vertical dimension erosion rate of the scour hole is consistent (see Figure 4), namely the formation process of morphology obviously presents three-dimensional characteristics. It is easy to understand that the morphology evolution of the state of equilibrium is also relatively slower. When $\theta \neq 90°$, the longitudinal dimension erosion rate of the scour hole is greater than that of the vertical dimension erosion rate; in other words, scour is mainly based on longitudinal dimension erosion, so the morphology evolution of the scour hole is relatively faster.

4.2. The Morphology and the Local Bed Shear Stress of Scour Hole

Under the clean-water scour condition, over time the scour range increases (the geometric size of the scour hole increases gradually over time.), the size of the horseshoe vortex increases, the intensity of the eddy decreases, and the shear stress of the local riverbed decreases and reaches the scouring equilibrium at $\tau_b < \tau_c$ [19,30]. The ratio adjustment law of τ_b/τ_c indicated that in the whole process of evolution, there was a significant difference in the rate of local bed shear stress decline, which showed that the rate of decline at the initial stage of scouring was fast, then the rate of decline was slow, which was clearly shown in Figure 8. Selamoğlu et al. [31] also reported the decreasing trend of shear stress evolution of riverbed in the scour hole, believing that a rapid decrease in the initial stage of the scour was followed by largely maintaining constant. Bouratsis et al. [11] also observed a similar phenomenon and pointed out that during the initial stage of the scour, the reduction of horseshoe vortex strength resulted in the reduction of local bed shear stress; In the development stage, the shear stress of the riverbed is still larger and lasts for a long time.

According to the evolution characteristics of scour hole morphology and the evolution trend of local bed shear stress, it is concluded that:

1. at $\tau_b > \tau_c$ stage: It is shown that the local shear stress is very large at the initial stage of local scour, which leads to rapid evolution of scour hole morphology. At the same time, the local shear stress decreases rapidly as the scour range expands rapidly.
2. at $\tau_b = \tau_c$ stage: It indicates that at the development stage of scouring, with the increase of scouring hole morphology, the local bed shear stress gradually approximates to the critical bed shear stress, thus causing the development rate of scour hole morphology to slow down and the local bed shear stress decrease to slow down.
3. at $\tau_b < \tau_c$ stage: The local shear stress is far less than the critical bed shear stress, which causes the local scour to slowly evolve into the scour equilibrium state. The evolution rate of scour hole morphology and the decrease rate of local bed shear stress are close to constant.

To sum up, during the whole process of scour evolution, the riverbed shear stress of scour hole decreases at two different rates. When the local scour depth reaches about 30% of the total scour depth, the riverbed shear stress of scour hole decreases quickly. To the scour equilibrium state, it slowly decreases and tends to constant. At each stage of scour evolution, local riverbed shear stress presents three typical characteristics respectively: $\tau_b > \tau_c$, $\tau_b = \tau_c$ and $\tau_b < \tau_c$.

5. Conclusions

Based on the flume experiment, this paper discusses the influence of spur dike alignment on the evolution characteristics of the scour hole around it.

Under the clean-water scour condition, the scour depth, the plane area and volume of scour hole, and evolution duration all present power function increases, and the scour hole profile area also accords with this function relation. Similar to scour depth evolution law, the plane area and volume of the scour hole also presents a power function over time. This characteristic has nothing to do with the alignment of the spur dike. However, the longitudinal dimension and vertical dimension erosion rate of the scour hole is affected significantly.

During the whole process of scour evolution and the development stage of the local scour, the local river bed shear stress of the scour hole decreases quickly. To the scour equilibrium state, it slowly decreases and tends to constant. Local scour evolution has three typical stages, and the local bed shear stress presents three characteristics respectively: $\tau_b > \tau_c$, $\tau_b = \tau_c$ and $\tau_b < \tau_c$.

During the whole evolution process, the scour hole-volume has a proportional constant with the product of the plane area and the maximum scour depth, the profile morphology of scouring hole on the upstream and downstream of spur dike remains approximately constant. With the change of the alignment angle, the scour range of the upstream and downstream of the spur dike is distinct and close to the state of scour balance. The length of the geometric characteristic is close to the golden section feature.

Author Contributions: L.Z., X.Z. conceived the experiments; B.W. designed and performed the experiments; J.C. analyzed the data and drew pictures; L.Z. and H.W. analysis and the interpretation of the results, wrote the paper.

Funding: This work was funded by the National Natural Science Foundation of China (51709116); Fundamental Research Funds for the Central Universities 2017B12214; Funding for the open project fund of Key Laboratory of the Yellow River sediment of the Ministry of Water Resources (201805).

Conflicts of Interest: The authors declare no conflicts of interest.

Nomenclature

L	The length of the spur dike
L'	Effective length of spur dike
d_{50}	The median particle size of sediment
h	Flow depth
θ	The alignment angle of the spur dike
U	Flow velocity
U_c	The threshold velocity of sediment
t	Any time of local scour evolution
T	The terminal time of scour evolution
d_{st}, d_{se}	Respectively are local scour depth at any time and at the scour equilibrium
A_{st}, A_{se}	Respectively are the plane area of local scour hole at any time and at the time of scour equilibrium
V_{st}, V_{se}	Respectively are local scour hole-volume at any time and at the time of scour equilibrium
A_0	The initial area of the local main vortex of the scour hole
A_{c-st}, A_{c-se}	Respectively are scour hole profile area at any time and at the time of scour equilibrium
τ_0	The flow shear stress of upstream spur dike
τ_b, τ_c	Respectively are the local bed shear stress of scour hole and the bed critical shear stress
u_*, u_{*cr}	Respectively are the frictional velocity of sediment particles and the critical frictional velocity
R_{1-st}, R_{3-st}	Respectively are the geometric characteristic length of the scour hole in different directions in each evolution stage

m, k	The index of describing the evolution of the scour plane area and volume
p, q	The index of describing the difference characteristic of longitudinal and vertical dimension evolution of the scour hole
$R_{1\text{-}st}, R_{3\text{-}st}$	Respectively are the plane geometric length of scour hole at upstream and downstream of the spur dike at any time of evolution

References

1. Melville, B.W.; Chiew, Y.M. Time scale for local scour at bridge piers. *J. Hydraul. Eng.* **1999**, *125*, 59–65. [CrossRef]
2. Oliveto, G.; Hager, W.H. Temporal evolution of clear-water pier and abutment scour. *J. Hydraul. Eng.* **2002**, *128*, 811–820. [CrossRef]
3. Ettema, R.; Muste, M. Scale effects in flume experiments on flow around a spur dike in flatbed channel. *J. Hydraul. Eng.* **2004**, *130*, 635–646. [CrossRef]
4. Cheng, N.S.; Chiew, Y.M.; Chen, X. Scaling analysis of pier-scouring processes. *J. Eng. Mech.* **2016**, *142*, 06016005. [CrossRef]
5. Lai, J.S.; Chang, W.Y.; Yen, C.L. Maximum local scour depth at bridge piers under unsteady flow. *J. Hydraul. Eng.* **2009**, *135*, 609–614. [CrossRef]
6. Diab, R.; Link, O.; Zanke, U. Geometry of developing and equilibrium scour holes at bridge piers in gravel. *Can. J. Civ. Eng.* **2010**, *37*, 544–552. [CrossRef]
7. Kuhnle, R.A.; Alonso, C.V.; Shields, F.D. Geometry of scour holes associated with 90 spur dikes. *J. Hydraul. Eng.* **1999**, *125*, 972–978. [CrossRef]
8. Kuhnle, R.A.; Alonso, C.V.; Shields, F.D. Local scour associated with angled spur dikes. *J. Hydraul. Eng.* **2002**, *128*, 1087–1093. [CrossRef]
9. Oscar, L.; Pfleger, F.; Zanke, U. Characteristics of developing scour-holes at a sand-embedded cylinder. *Int. J. Sediment Res.* **2008**, *23*, 258–266.
10. Bouratsis, P.; Diplas, P.; Dancey, C.L.; Apsilidis, N. Quantitative spatio-temporal characterization of scour at the base of a cylinder. *Water* **2017**, *9*, 227. [CrossRef]
11. Gaudio, R.; Tafarojnoruz, A.; Calomino, F. Combined flow-altering countermeasures against bridge pier scour. *J. Hydraul. Res.* **2012**, *50*, 35–43. [CrossRef]
12. Williams, P.D. Scale Effects on Design Estimation of Scour Depths at Piers. Master's Thesis, University of Windsor, Windsor, ON, Canada, April 2014.
13. Zhang, L.; Wang, P.; Yang, W.; Zuo, W.; Gu, X.; Yang, X. Geometric characteristics of spur dike scour under clear-water scour conditions. *Water* **2018**, *10*, 680. [CrossRef]
14. Kothyari, U.C.; Garde, R.C.J.; Ranga Raju, K.G. Temporal variation of scour around circular bridge piers. *J. Hydraul. Eng.* **1992**, *118*, 1091–1106. [CrossRef]
15. Kothyari, U.C.; Kumar, A. Temporal variation of scour around circular compound piers. *J. Hydraul. Eng.* **2012**, *138*, 945–957. [CrossRef]
16. Kumar, A.; Kothyari, U.C.; Ranga Raju, K.G. Flow structure and scour around circular compound bridge piers—A review. *J. Hydro-Environ. Res.* **2012**, *6*, 251–265. [CrossRef]
17. Pandey, M.; Sharma, P.K.; Ahmad, Z.; Karna, N. Maximum scour depth around bridge pier in gravel bed streams. *Nat. Hazards* **2017**, *91*, 819–836. [CrossRef]
18. Dey, S. Time-variation of scour in the vicinity of circular piers. *Proc. Inst. Civ. Eng. Water Marit. Energy* **1999**, *136*, 67–75. [CrossRef]
19. Dey, S.; Barbhuiya, A.K. Time variation of scour at abutments. *J. Hydraul. Eng.* **2005**, *131*, 11–23. [CrossRef]
20. Yanmaz, A.M.; Kose, O. A semi-empirical model for clear-water scour evolution at bridge abutments. *J. Hydraul. Res.* **2009**, *47*, 110–118.
21. Melville, B.W.; Raudkivi, A.J. Flow characteristics in local scour at bridge piers. *J. Hydraul. Res.* **1977**, *15*, 373–380. [CrossRef]
22. Barbhuiya, A.K. Clear Water Scour at Abutments. Ph.D. Thesis, Indian Institute of Technology, Kharagpur, India, 2003.
23. Zhang, L.; Sun, K.Z.; Xu, D.P. Morphological evolution of spur dike local scour hole and the scour balance critical condition. *J. Hydraul. Eng.* **2017**, *48*, 545–550.

24. Bateman, A.; Fernández, M.; Parker, G. Morphodynamic model to predict temporal evolution of local scour in bridge piers. In Proceedings of the 4th IAHR Symposium on River, Coastal and Estuarine Morphodynamics (RCEM 2005), Urbana, IL, USA, 4–7 October 2005; pp. 911–920.

25. Kirkil, G.; Constantinescu, G. Flow and turbulence structure around an in-stream rectangular cylinder with scour hole. *Water Resour. Res.* **2010**, *46*, W11549. [CrossRef]

26. Melville, B.W.; Coleman, S.E. *Bridge Scour*; Water Resources Publication, LLC: Littleton, CO, USA, 2000.

27. Whitehouse, R.J.S. *Scour at Marine Structures: A Manual for Engineers and Scientists*; Research Report SR417; HR Wallingford Limited: Wallingford, UK, 1997.

28. Masjedi, A.; Bejestan, M.S.; Moradi, A. Experimental study on the time development of local scour at a spur dike in a 180 flume bend. *J. Food Agric. Environ.* **2010**, *8*, 904–907.

29. Dodaro, G.; Tafarojnoruz, A.; Sciortino, G.; Adduce, C.; Calomino, F.; Gaudio, R. Modified Einstein sediment transport method to simulate the local scour evolution downstream of a rigid bed. *J. Hydraul. Eng.* **2016**, *142*, 04016041. [CrossRef]

30. Yilmaz, M.; Yanmaz, A.M.; Koken, M. Clear-water scour evolution at dual bridge piers. *Can. J. Civ. Eng.* **2017**, *44*, 298–307. [CrossRef]

31. Selamoğlu, M. Modeling Temporal Variation of Scouring at Dual Bridge Piers. Ph.D. Thesis, Middle East Technical University, Ankara, Turkey, June 2015.

water

MDPI

Article

The Logarithmic Law of the Wall in Flows over Mobile Lattice-Arranged Granular Beds

Federica Antico [1], Ana M. Ricardo [2] and Rui M. L. Ferreira [1,2,*]

[1]   Instituto Superior Técnico, Universidade de Lisboa, 1049-001 Lisbon, Portugal;
      federica.antico@tecnico.ulisboa.pt
[2]   CERIS—Civil Engineering Research and Innovation for Sustainability, 1049-001 Lisbon, Portugal;
      ana.ricardo@tecnico.ulisboa.pt
*   Correspondence: ruimferreira@tecnico.ulisboa.pt

Received: 4 April 2019; Accepted: 23 May 2019; Published: 4 June 2019

Abstract: The purpose of the present paper is to provide further insights on the definition of the parameters of the log-law in open-channel flows with rough mobile granular beds. Emphasis is placed in the study of flows over cohesionless granular beds composed of monosized spherical particles in simple lattice arrangements. Potentially influencing factors such as grain size distribution, grain shape and density or cohesion are not addressed in this study. This allows for a preliminary discussion of the amount of complexity needed to obtain the log-law features observed in more realistic open-channel flows. Data collection included instantaneous streamwise and bed-normal flow velocities, acquired with a two-dimensional and two-component (2D2C) Particle Image Velocimetry (PIV) system. The issue of the non uniqueness of the definition of the parameters of the log-law is addressed by testing several hypotheses. In what concerns the von Kármán parameter, κ, it is considered as flow-independent or flow-dependent (a fitting parameter). As for the geometric roughness scale, k_s, it results from a best fit procedure or is computed from a roughness function. In the latter case, the parameter B is imposed as 8.5 or is calculated from the best fit estimate. The analysis of the results reveals that a flow dependent von Kármán parameter, lower than the constant $\kappa = 0.40$, should be preferred. Forcing $\kappa = 0.40$ leads to non-physical values of k_s and would imply extending the inner layer up about 50% of the flow depth which is physically difficult to explain. Considering a flow dependent von Kármán parameter allows for coherent explanations for the values of the remaining parameters (the geometric roughness scale k_s, the displacement height Δ, the roughness height z_0). In particular, for the same transport rate, the roughness height obtained in a natural sediment bed is much greater than in the case of bed made of monosized glass spheres, underlining the influence of the bed surface complexity (texture and self-organized bed forms, in the water-worked cases) on the definition of the log-law parameters.

Keywords: logarithmic law of the wall; von Kármán parameter κ; bedload; granular beds; drag-reducing flows

1. Introduction

The classical idealization of flows over smooth and rough boundaries, successfully extended to mobile boundaries, comprises a logarithmic distribution of the longitudinal velocity in the wall-normal direction. In open channel flows, this log-law should be valid in the overlapping layer between inner and outer flow regions (see, e.g., [1]) when (i) gradients in the longitudinal direction are small, in particular the pressure gradient; (ii) the channel aspect ratio is high so that the mean flow far from the banks or side walls is two-dimensional in the vertical plane and (iii) the relative submersion is high. The log-law can be written as

$$\frac{\langle \bar{u} \rangle}{u_*} = \frac{1}{\kappa} \ln \left(\frac{z'}{z_0} \right), \tag{1}$$

where $\langle \bar{u} \rangle$ is the space and time-averaged longitudinal velocity, $z' = Z - Z_c$ is the vertical coordinate above the zero of the log-law (Z_0), Z is the vertical coordinate above and arbitrary datum, z_0 is the bed's characteristic roughness height (such that $\langle \bar{u} \rangle (z' = z_0) = 0$), κ is the von Kármán parameter and $u_* = \sqrt{\tau_0/\rho^{(w)}}$ is the friction velocity (the kinematic scale for both inner and outer flow variables), where τ_0 is the wall shear stress and $\rho^{(w)}$ is the fluid density.

For hydraulically rough boundaries, the idealization of inner and outer flow regions may not be sufficient to describe the complexity of the flow in the vicinity of the roughness elements. Several conceptual models have been proposed to include in the description of flow stratification a near-bed layer where bed micro-topography determines mean flow features. For the sake of consistency with the early work of some of the authors, the idealization of Ferreira et al. [2], depicted in Figure 1, is followed in this text. In region (A), the turbulent flow is influenced by the free-surface. In the inner region (B), the flow is affected by the characteristics of the rough wall, directly in the lowermost layers and indirectly, through u_*, in the uppermost layers. The dominant characteristic length scale in region (B) is k_s, mostly influenced by the diameter of the granular material and its superficial arrangement (the bed micro-topography). The characteristic length scale should be of the order of magnitude of the bed amplitude, δ (the wall-normal distance between the planes of the troughs and of the crests) but should be dependent on the type of granular bed and on the Shields parameter [2].

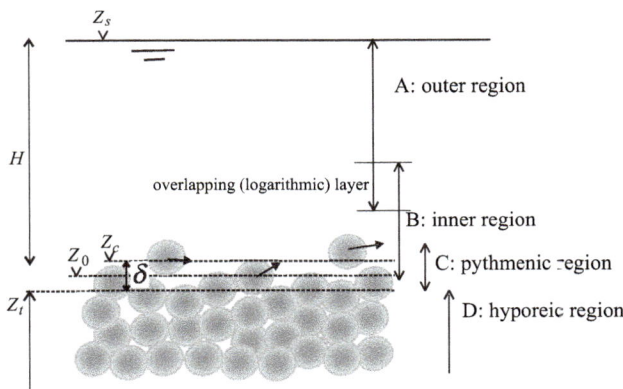

Figure 1. Idealized physical system. Z_s is the elevation of the free-surface, Z_c and Z_t are the space-averaged elevations of the planes of the crests and of the bed troughs, respectively. The bed amplitude is $\delta = Z_c - Z_t$. All elevations are relative to an arbitrary datum.

The lower boundary of the inner region is set at the elevation of the zero of the log-law, Z_0, located $|\Delta|$ above (or below) the boundary zero. Δ is known as the displacement height and can be negative if the zero of the log-law is above the boundary zero. It is noted that the boundary zero is arbitrary. In this text, it is assumed to be contained in the plane of the mean elevation of the crests. Assuming wall similarity in the sense of Townsend [1], i.e., balance of production and dissipation rates of turbulent kinetic energy and near-constant Reynolds shear stresses, in the overlapping layer between the inner and outer regions, the wall-normal profile of the longitudinal velocity is expressed by Equation (1). In the pythmenic region (C), whose upper boundary can be located above the plane of the crests if the bedload discharge is large, the flow is three-dimensional and mainly determined by the bed micro-topography and granular movement. Below the elevation of the troughs, Z_t, in the hyporheic region (D), the flow may be described by Darcy or Brinkman models [3].

The universality of the von Kármán parameter in open-channel flows has long been discussed in the literature. For instance, the assumption of a von Kármán parameter different from a constant value (usually between $\kappa = 0.40$ and $\kappa = 0.41$) is discussed in case of flows with low submergence or in the presence of suspended or bedload transport by [4]. The influence of bed mobility on the value of the von Kármán parameter has been reported for more than 20 years in several experimental studies, normally featuring a reduction of the value of κ (e.g., [5]). Gaudio et al. [4] have reviewed and published experimental results and have proposed the non-universality of κ in flows over sediment beds. These authors have shown that, depending on flow characteristics, bed configuration and bedload transport rate, the von Kármán parameter may assume values ranging between 0.29 and 0.39. Gaudio et al. [6], analyzing the velocity field in a flume with fixed bed and a sediment supply of coarse sand (d_{50} = 1 mm), observed a range of κ between 0.3 and 0.4 with a marked reduction of the value of κ with increased sediment concentration. These results were obtained assuming a bed Nikuradse sand equivalent roughness of $k_s = 2d_{50}$. In this case, the von Kármán parameter was calculated as $\kappa = \frac{u_*}{a}$, where a is the slope of the semi-log velocity profiles in the logarithmic layer of the inner region. A zero value of zero-plane displacement was estimated following the procedure presented in Koll [7].

Recently, Hanmaiahgari et al. [8] computed the parameters of the log-law in flows over hydraulically rough mobile beds subjected to different conditions: immobile, weakly mobile and temporally varying mobile bed conditions with different stages of bedform development. The von Kármán parameter κ was evaluated from the logarithmic law of the wall with the zero of bed-normal axis, z, set at the plane of the crests of the particles (Z_c). The displacement height Δ was considered equal to $0.35d_{50}$. Their experimental results showed a decrease of κ due to increase of the thickness of roughness sublayer as the mobility of bedforms increases: the von Kármán parameter κ varied between 0.27 in the case of mobile bedforms and 0.41 for immobile bed conditions. Working with bimodal mixtures of sand and gravel (simulating a natural rough bed), the influence of bedload transport on the values of the von Kármán parameter κ has been investigated by Ferreira et al. [2].

Contrary to the experimental findings of the authors discussed above, Ferreira et al. [2] observed that the vertical profiles of the longitudinal velocity, in case of water-worked beds of sand-gravel mixtures, could be fitted to a log-law either with a flow independent (κ = 0.40) or a flow-dependent von Kármán parameter. They proposed that the value of κ depends on the interpretation of the log-law parameters such as boundary zero, the geometric roughness scale k_s and the displacement height Δ. They found out that the location of the zero-plane displacement for the log-law could not remain constant and should increase with the increase of the bedload discharge, if κ is set to 0.40. On the contrary, assuming the von Kármán parameter as flow dependent, they showed a decrease in terms of displacement height Δ at the onset of generalized bed load transport, together with a drop on the value of κ. In both cases, the zero for the log-law was below the plane of the sediment crests. In the wake of Ferreira et al. [2], Ferreira [9] provided a similarity framework to discuss the nature of κ in flows over weakly mobile gravel-sand beds and explored a theoretical model for κ as a function of turbulence parameters. There are three hypotheses of interpretation of his theoretical framework for the cases of: no-similarity, complete similarity or incomplete similarity in the non-dimensional parameters describing bed composition and bed mobility.

In the first case (no similarity), Ferreira [9] argued that the vertical distribution of the longitudinal mean velocity would not be logarithmic. If complete similarity exists, on the contrary, all flows characterized by rough mobile beds should be identified by a normalised shear rate in the overlapping layer independent of bed composition and bed mobility and by a constant κ, not necessarily equal to 0.40, independent from Reynolds and Shields numbers. This is the case supported by the laboratory results on the structure of turbulence for flows over rough mobile beds (gravel-sand mixtures) reported in [9]. The author defends that changes in the structure of turbulence in the inner region do not seem to depend continuously on the Shields number and do not imply a reduction of κ, in case of mobile bed conditions, with respect to κ = 0.40. In the case of incomplete similarity, Ferreira [9] proposed a dependency of the constant shear rate on the overlapping layer on bed mobility; κ was considered a function of bed composition and Shields number. Both complete and incomplete similarity criteria required a joint discussion of κ and of the displacement height Δ. Ferreira [9] and Ferreira et al. [2] argued that different choices of Δ may lead to different values of κ without physical justification.

Bearing in mind that most of the studies above featured sediment beds with different degrees of complexity, the key objective of this paper is to provide insights on the definition of the parameters of the log-law in rough mobile granular beds when complexity is reduced to minimum. Factors influencing surface roughness such as grain size distribution, grain shape and density or cohesion are not addressed in this study. Data collection hence took place in flows over cohesionless granular beds composed by monosized spheres—5 mm glass beads—arranged in simple lattices. These are juxtapositions of close packing arrangements with body-centered cubic arrangements. These experimental tests are meant to clarify the description and interpretation of the parameters of the log-law in rough mobile beds casting aside the complexity introduced by working with natural sediment.

The issue of the nonuniqueness of the definition of the log-law parameters is addressed by analyzing and discussing the experimental results under different scenarios: scenarios 1 and 2 consider κ constant and equal to 0.41 and different definitions of geometric roughness scale k_s; scenarios 3 and 4 retake the definition of k_s reported in the first and second scenario respectively, but the von Kármán parameter is, in this case, considered flow-dependent. The procedures adopted to interpret the laboratorial data follow those of Ferreira et al. [2]. Issues of universality and uniqueness in the definition of the parameters of the log-law are still important topics of research not only for its intrinsic value—advancing fundamental knowledge—but also because of its direct impacts on the quality of the predicting mathematical modelling tools. The log-law is frequently used as a wall function in Computational Fluid Dynamics (see reviews in e.g., [10] or [11]), in the context of Reynolds-averaged Navier-Stokes equations (RANS) modelling (e.g., [12], detached eddy simulation (DES) or even large eddy simulation (LES) modelling ([13,14]) or integral Navier-Stokes (NS) equation modelling ([15]). Furthermore, depth-averaged hydrodynamic and sediment transport models use the parameters of the log-law to estimate the wall shear stress (e.g., [16–18]). Improving the accuracy of predictive models for flows over mobile boundaries thus requires investment in the definition of the parameters of the log-law.

2. Laboratory Facilities, Instrumentation and Procedures

Laboratory tests were performed in a 12.5 m long and 40.5 cm wide prismatic channel, recirculating water and sediment through independent circuits, installed at the Laboratory of Hydraulics and Environment of Instituto Superior Técnico, Lisbon (Figure 2a). The channel has side glass-walls enabling flow visualization and laser illumination. The flume bed was divided in two main reaches:

- a fixed-bed reach comprising 1.5 m of large boulders (50 mm average diameter), followed by 3.0 m of smooth bottom (PVC) and 2.5 m of one layer of glued spherical glass beads (5.0 mm diameter) to ensure the development of a rough-wall boundary layer (Figure 2a,b);
- a mobile reach 4.0 m long and 2.5 cm deep filled with 5.0 mm diameter glass beads, with density $\rho_s = 2490 \, \text{kg/m}^3$, packed (with some vibration) to a void fraction of 0.325, expressing the mixed nature of the lattice arrangement (face-centered and body centered), seen in Figure 2c).

Flow is nearly uniform. In all tests, the energy slope may deviate less than 20% from the channel slope. Since the phenomena of interest to this paper take place in the inner flow layer, this fact is considered not important. Data collected for four experimental tests are discussed in this text: test 1 was performed under sub-threshold conditions (no particles moving), while tests 2, 3 and 4 are respectively characterized by bedload rate of 6.23, 21.12 and 28.72 beads/s.

Free surface elevation and bed level were measured with 0.5 mm precision point gage in five transversal sections of the flume and in three lateral positions per cross-section.

The water discharge at the flume inlet was controlled by two digital flowmeters respectively installed on the main water recirculating pump and on the secondary pump recirculating water and sediments. The inlet of the sediments was at $x = 3.0$ m, measured from the main water inlet. The secondary pump provided a constant flow rate of 0.0020 m^3/s. The flowrate released by the main pump was adjusted depending on the test to obtain the total flow rates reported, among other main characterizing variables, in Table 1.

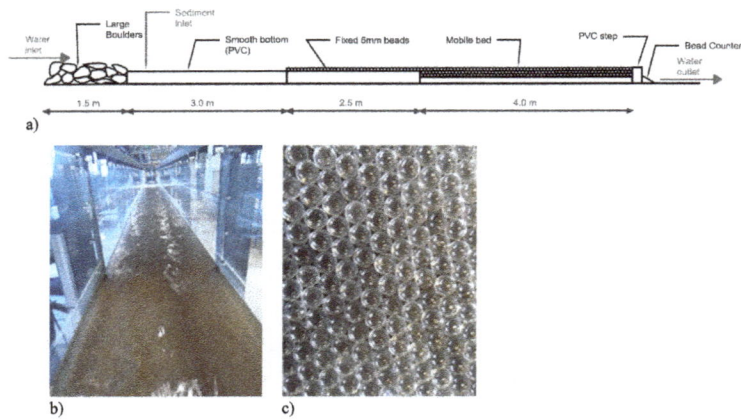

Figure 2. (**a**) Scheme of the complete flume setup; (**b**) general view of the flow over the mobile-bed reach; (**c**) granular bed, prior to water working, showing the simple lattice arrangement.

Table 1. Characterization of the mean flow.

Test	Q (m^3/s)	H (m)	i_b (-)	U (m/s)	$u_*^{(1)}$ (m/s)	$\tau_b^{(1)}$ (Pa)	$u_*^{(2)}$ (m/s)	$\tau_b^{(2)}$ (Pa)
1	0.0150	0.071	0.00317	0.518	0.041	1.64	0.041	1.68
2	0.0167	0.068	0.00456	0.602	0.048	2.29	0.048	2.26
3	0.0208	0.074	0.00623	0.691	0.058	3.33	0.056	3.15
4	0.0214	0.070	0.00714	0.757	0.060	3.63	0.061	3.73

The other variables in Table 1 are: the mean flow depth, H, defined as the wall-normal distance between Z_s and Z_c (see Figure 1); the slope of the flume, i_b; the depth-averaged mean longitudinal velocity, U; the friction velocity and bed shear stress computed from the equation of conservation of momentum in x direction, assuming uniform flow, respectively $u_*^{(1)} = \sqrt{\dfrac{\tau_b^{(1)}}{\rho_w}}$ and $\tau_b^{(1)} = \gamma_w R_h i_b$, where ρ_w is the water density, $\gamma_w = g\rho_w$ is the volumetric weight of the water and R_h is the hydraulic radius; and the friction velocity and the bed shear stress calculated from the wall-normal profile of the wall-normal component of the turbulence kinetic energy, respectively $u_*^{(2)}$ and $\tau_b^{(2)} = \rho_w u_*^{(2)^2}$. The friction velocity $u_*^{(2)}$ was computed as the maximum of the profile $C\sqrt{\overline{(w'w')}}(z)$. Coefficient C is not universal; Nezu et al. [19], for instance, assumed $C \approx 0.83$. If the relative importance of the terms of the turbulence kinetic energy is the same as in Nezu et al. [19], the coefficient proposed by Soulsby and Dyer [20] would be $C = 0.82$ and the coefficient in Stapleton and Huntley [21] would be $C = 0.80$. In this study, a constant value of $C = 1.00$ was employed as it minimized the mean square error between $u_*^{(2)}$ and $u_*^{(1)}$.

Table 2 shows the values of the relevant non-dimensional parameters: Froude number, $Fr = \dfrac{U}{\sqrt{gh}}$, Shields parameter, $\theta = \dfrac{u*^{(2)}}{(s-1)gd}$, Reynolds number of the mean flow, $Re = \dfrac{Uh}{\nu^{(w)}}$ (where $\nu^{(w)}$ is the fluid's kinematic vicosity), bed Reynolds number, $Re* = \dfrac{u_*^{(2)}d}{\nu^{(w)}}$ and non-dimensional bedload discharge, $\Phi = \dfrac{q_s}{\sqrt{(s-1)gd^3}}$. The bedload discharge rate q_s was determined by using the particle counting system described in detail in Mendes et al. [22]. This device was installed at the downstream end of the mobile bed reach of the flume. The pressure variations produced by the impacts of the sediment particles falling on the membrane boxes (placed over the entire width of the channel) were registered and analyzed. The accumulated count registered once flow and sediments' equilibrium conditions were achieved was then translated into solid flow discharge per each test. The bedload rate was evaluated as $q_s = \frac{Vn}{b}$, where $V = 6.545 \times 10^{-8}$ m^3 is the volume of the glass particles, n is the number of beads impacting on the bead counter per second and b is the channel's width.

Table 2. Non-dimensional parameters characterizing the mean flow.

Test	Fr	Re	Re$_*$	θ	Φ
1	0.62	41,405	227	0.023	0.0000
2	0.73	46,057	268	0.030	0.0007
3	0.81	57,571	323	0.042	0.0025
4	0.92	58,999	337	0.050	0.0034

The instantaneous flow velocity (longitudinal u and vertical w) was measured with a two-dimensional and two-component (2D2C) Particle Image Velocimetry (PIV) system, in three different longitudinal positions: 10.2 cm, 20.4 cm and 30.6 cm from the right channel sidewall, where the position at 20.4 cm represents the center longitudinal section of the flume. The observation window for flow velocity measurements was placed at 2.5 m from the beginning of the mobile bed reach, covering a length that comprises between 6 cm and 12 cm, depending on the test, and covers the entire flow depth. An acetate sheet was placed on the water surface to ensure optical stability and absence of laser sheet reflections. The PIV system consisted of an 8 bit 1600 × 1200 px^2 CCD camera and a double-cavity solid state laser with pulse energy of 30 mJ at wavelength of 532 nm. The system was operated at 15 Hz with a time between pulses within the range from 380 μs to 500 μs. Polyurethane particles with mean diameter of 50 mm in a range from 30 to 70 mm and specific density of 1.31 g/cm^3 were used as solid targets. Such tracer particles have a cut-off frequency of approximately 2500 Hz for a significance level of 0.95 (Ferreira and Aleixo [23]). Hence, turbulence with frequencies lower than 2500 Hz is likely to be well-represented by the motion of the tracer particles. Since the PIV was operated at 15 Hz, the Nyquist frequency of the time series is 7.5 Hz, much smaller than the cut-off frequency. Hence, the employed seeding particles do not constitute an extra limitation to the time resolution of the PIV.

The duration of each PIV time series was 5 min of consecutive data corresponding to 4500 image couples per each measurement position. DynamicStudio software (version 3.41, Dantec Dynamics®, Skovlunde, Denmark) allowed for processing image pairs with the adaptive correlation algorithm. The initial interrogation area was of 128 × 128 px^2, while the final was of 16 × 16 px^2, with an overlap of 50%.

3. Data Analysis and Results

3.1. PIV Post-Processing

The images acquired by the PIV system were post-processed by masking the areas in the field of view not occupied by fluid: the band above the free surface (identified as the trace of the laser sheet on the fluid surface) and the region occupied by the bed particles (both immobile and mobile particles). A specially designed algorithm based on thresholding and median and edge filters was developed to automatically detect bed particles edge contours (see details in [24]). A fixed mask was applied to the time-averaged free surface since its oscillations were small.

Image masking allowed for computing the space–time porosity $\phi_{VT}(x_i, t)$ (Nikora et al. [25]). Every image domain can be divided in two parts, one occupied by fluid and one by solid (sediment), identified by a clipping or distribution function, γ, set to 1 in the fluid and 0, otherwise. According to [25], the space–time porosity $\phi_{VT}(x_i, t)$ is defined as:

$$\phi_{VT}(x_i, t) = \frac{1}{T_0} \frac{1}{V_0} \int_{T_0} \int_{V_0} \gamma(x_i + \xi_i, t + \tau) dV d\tau, \tag{2}$$

where T_0 is the averaging period, V_0 is the spatial averaging domain, and γ is the clipping function. The integration domain is centered at position x_i and t. A local coordinate system in space and time, respectively ξ_i and τ, is used for integration.

The velocity profiles discussed in Section 3.2 are the average velocity values of the three longitudinal sections in which PIV acquisition was performed. The boundary zero was set, for each vertical plane, with the criteria presented above. An interpolation in an evenly space grid of discrete increments $z = 0.0006$ m between the zero and the maximum free surface level was applied before averaging.

3.2. Calculation of the Parameters of the Log-Law

Ferreira et al. [2] proposed three possible scenarios to understand their laboratorial data associated with different definitions of the parameters and scale of the log-law. The bed material used in their experiments was a gravel-sand or a gravel mixture depending on the specific test. The location of the boundary zero was set at the elevation of the lowest troughs in scenario (s1) and scenario (s3) and at the plane of the higher crests in scenario (s2).

In the present experiments, the reference zero is set at the level of the crests of the roughness elements. For practical purposes, and given that particles transported as bedload induce a layer where ϕ_{VT} is less than one, the reference zero was defined as the level of the wall-normal scale corresponding to $\phi_{VT} = 0.95$. The double-averaged longitudinal velocity profiles and the space–time porosities obtained for the central section of the flume are shown in Figure 3.

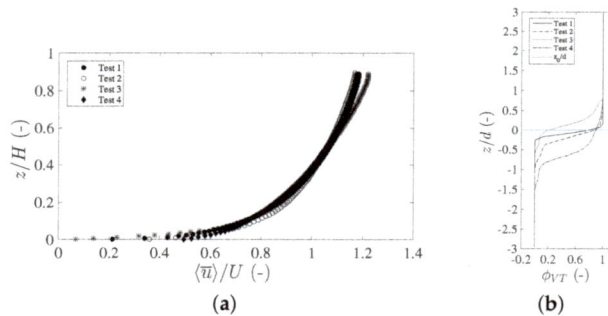

Figure 3. Double-averaged longitudinal velocity profiles (**a**) and space–time porosity $\phi_{VT}(x_i, t)$ (**b**). The reference zero in plot (**a**) corresponds to the elevation of the particle crests (defined in the text). H is the flow depth measured from this zero. The reference zero in plot (**b**) corresponds to the initial elevation of the particle crests for Test 1 (no bedload transport).

Following the criteria advanced by Ferreira et al. [2], the results of the present experimental study are interpreted in accordance with four different hypotheses:

1. the von Kármán parameter is considered flow independent ($\kappa = 0.405$), the geometric roughness scale k_s and the constant B are subjected to a best fit procedure.
2. the von Kármán parameter is considered flow independent ($\kappa = 0.405$), the constant B is 8.5 and the roughness scale k_s is calculated from a roughness function.
3. the von Kármán parameter is assumed not universal but a fitting parameter, the geometric roughness scale k_s and the constant B are subjected to a best fit procedure.
4. the von Kármán parameter is assumed not universal but a fitting parameter and, as in scenario 2, the constant B is imposed equal to 8.5 and the roughness scale k_s is calculated from a roughness function.

In scenario 1, the displacement height Δ, i.e., the elevation of the zero-plane for the logarithmic law, is derived from:

$$\left\{ \frac{d}{dz} \frac{\langle \overline{u} \rangle}{u_*} \right\}^{-1} = \kappa z - \kappa \Delta, \tag{3}$$

where $\kappa = 0.405$.

A linear regression on the values of Equation (3) renders the value of Δ from the origin of the regression line. Once the displacement height Δ is defined, the remaining parameters of the log-law (k_s and B) are retrieved from the best fit procedure of the log-law written in the form ([26]):

$$\frac{\langle \overline{u} \rangle}{u_*} = \frac{1}{\kappa} \ln \frac{z - \Delta}{k_s - \Delta} + \frac{\langle \overline{u} \rangle_I}{u_*}, \tag{4}$$

where $u_* \equiv u_*^{(2)}$ is the friction velocity (see Table 1) and $\langle \overline{u} \rangle_I$ is the velocity at the lower bound of the logarithmic layer.

The bounds of the regression analysis to determine k_s and B were adjusted to maximize the coefficient of determination r^2 while maintaining $\kappa = 0.405$.

In scenario 2, the displacement height Δ is retrieved with the same procedure of scenario 1. In this case, the log-law is written as [2]:

$$u = \frac{u_*}{\kappa} \ln (z - \Delta) - \frac{u_*}{\kappa} \ln (k_s) + u_* B, \tag{5}$$

with $B = 8.5$. Equation (5), which is in the form $Y = MX + A$, is fitted to the data with $M = \frac{u_*}{\kappa}$ and $A = u_* \left(B - \frac{1}{\kappa} \ln(k_s) \right)$ as fitting parameters. Again, the bounds of the regression were adjusted so that the von Kármán parameter κ approached 0.405. Finally, the scale of the roughness elements is computed from $\ln(k_s) = \kappa(8.5 - \frac{A}{u_*})$.

In scenario 3, the data of Equation (3) are fitted to a linear reach with no *a priori* considerations about the von Kármán parameter κ. The geometric scale of the roughness elements k_s is computed as in scenario 1.

Once the bounds of the regression on the values of Equations (3) and (4) are set in order to maximize the coefficient of determination r^2, the displacement height Δ and the von Kármán parameter κ are defined through Equation (3), while $B = \frac{\langle \overline{u} \rangle_I}{u_*}$, through Equation (4).

Scenario 4 differs from scenario 2 in as much as the von Kármán parameter κ is not considered universal. The value of κ determined as in scenario 3.

Finally, the roughness height, z_0, can be computed from k_s applying the following relation, valid for all four approaches:

$$\frac{z_0}{k_s} = e^{-\kappa B}, \tag{6}$$

where B, κ and k_s are the parameters found for each scenario. The values of the log-law parameters are displayed in Tables 3–5.

Table 3. Parameters describing the log-law for Scenario 1, where $\kappa = 0.405$.

Test	Δ (m)	k_s (m)	B (-)	z_0 (m)
1	0.0072	0.0336	14.7	8.7×10^{-5}
2	0.0020	0.0192	14.1	6.5×10^{-5}
3	0.0080	0.0420	14.5	1.2×10^{-4}
4	0.0037	0.0312	14.0	1.1×10^{-4}

Table 4. Parameters describing the log-law for Scenario 2, where $\kappa = 0.405$ and $B = 8.5$.

Test	Δ (m)	k_s (m)	z_0 (m)
1	0.0072	0.0021	6.7×10^{-5}
2	0.0020	0.0018	5.8×10^{-5}
3	0.0080	0.0029	9.4×10^{-4}
4	0.0037	0.0030	9.6×10^{-5}

Table 5. Parameters describing the log-law for Scenario 3, where κ is not considered universal.

Test	Δ (m)	k_s (m)	B (-)	κ (-)	z_0 (m)
1	-0.0001	0.0018	9.54	0.352	1.7×10^{-4}
2	-0.0002	0.0036	10.26	0.350	1.0×10^{-4}
3	-0.0005	0.0048	9.78	0.355	1.5×10^{-4}
4	-0.0007	0.0060	10.41	0.305	2.5×10^{-4}

3.3. Discussion of the Values of the Parameters of the Log-Law

The geometrical outcome of the best fit procedure from which the values of κ and Δ, shown in Tables 3–6, were derived can be seen in Figure 4. It is clear in the figures that there are at least two possible ways of fitting a linear model, if different reaches of the wall-normal coordinate are selected. Since the von Kármán parameter κ is the slope of the regression line and the displacement height Δ depends on the intercept, two possible values of the pair (κ, Δ) are derived from the same data set. One of the possible outcomes corresponds to scenarios 1 and 2, for which $\kappa = 0.405$ (Figure 4a,b, respectively). The other possibility corresponds to the flow dependent κ scenarios 3 and 4 (Figure 4c,d, respectively).

Table 6. Parameters describing the log-law for Scenario 4, where κ is not considered universal and $B = 8.5$.

Test	Δ (m)	k_s (m)	κ (-)	z_0 (m)
1	-0.0001	0.0041	0.352	2.1×10^{-4}
2	-0.0002	0.0030	0.350	1.5×10^{-4}
3	-0.0005	0.0062	0.355	3.0×10^{-4}
4	-0.0007	0.0074	0.305	5.5×10^{-4}

The bounds of the regression analysis represented in red in Figure 4 were set so as to maximize the coefficient of determination. Blue lines represent the bounds associated to the minimum admissible coefficient of determination. The latter is set as 98% of the maximum coefficient of determination.

As noted by Ferreira et al. [27] and Ferreira et al. [26], for rough mobile bed in the presence of bedload transport, the velocity profiles may not be self-similar due to bed mobility affecting the parameters of the log-law. This possibility has been discussed in detail by Ferreira [9] and formalized as incomplete similarity in the parameters that describe bed composition and mobility. This is explored in the present study through scenarios 3 and 4 for which the slope of the inverse shear rate is not the same across the tests (Figure 4) and is consistently less than 0.405.

From Equation (3), the displacement height Δ is computed as the intercept of the regression line divided by $-\kappa$. A small but negative displacement height is obtained for scenarios 3 and 4, corresponding to the flow-dependent κ. As seen in Tables 5 and 6, the zero of the log-law is just above the crests of the roughness elements and within the layer where bedload occurs. Positive values of Δ are attained for scenarios 1 and 2 (corresponding to $\kappa = 0.405$). The obtained values of Δ are of the order of magnitude of the diameter of the glass beads, which has not been reported in earlier studies.

The double-averaged wall-normal profiles of the longitudinal velocity are depicted in Figure 5 (scenario 1), Figure 6 (scenario 2), Figure 7 (scenario 3) and Figure 8 (scenario 4). The best fit lines resulting from the regression analysis and their lower and upper boundaries are also shown in these figures.

In open-channel flows, the overlapping region in which the log-law is defined occurs generally $z/h \leq 0.2$ [19]. In this region, the present data support a Von Kármán parameter κ lower than 0.405. On the other hand, the region characterized by $\kappa = 0.405$ is located well above the lower 20% of the total flow depth (around 35–80%, depending on the test).

In scenario 1, the values of the scale of the roughness elements k_s, coinciding with the lower bound of the linear reach, are considerably high—between 4 and 8 bead diameters above the reference zero. Parameter B exhibits also relatively high values, well above the classical $B = 8.5$ ([28]). In scenario 2, the values for the scale of roughness elements also appear implausible—0.36–0.60 bead diameters—since they are less than one bead diameter. The overlapping region between outer and inner region is also observed, as in scenario 1, rather high on the water column (35–80%).

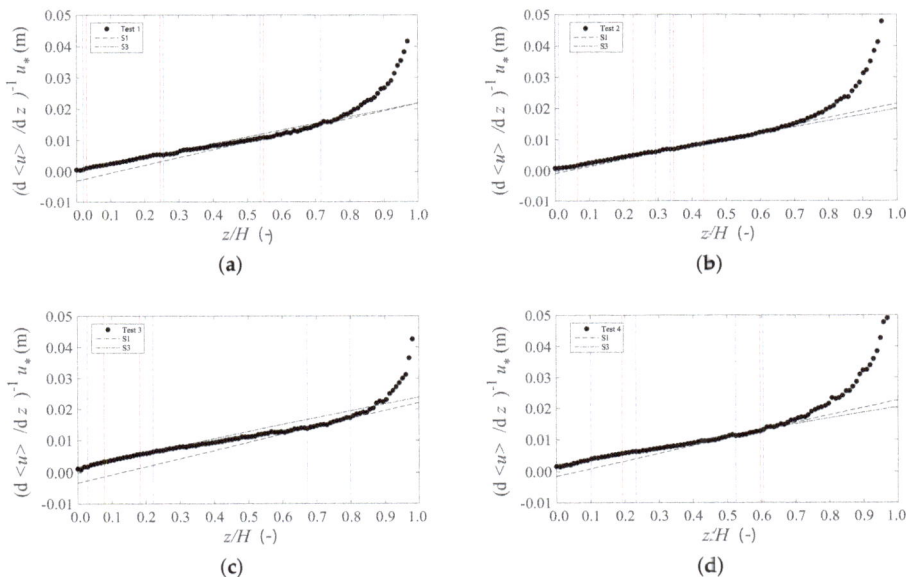

Figure 4. Shear rate and two-linear reaches identified respectively for Test 1 (**a**); Test 2 (**b**); Test 3 (**c**); Test 4 (**d**). The regression lines are represented by dashed and dotted lines (as identified in the legend). The bounds of the regression analysis that maximize the determination coefficient are marked with vertical red dashed lines. Blue lines represent the bounds associated with the minimum admissible coefficient of determination—98% of the maximum coefficient of determination.

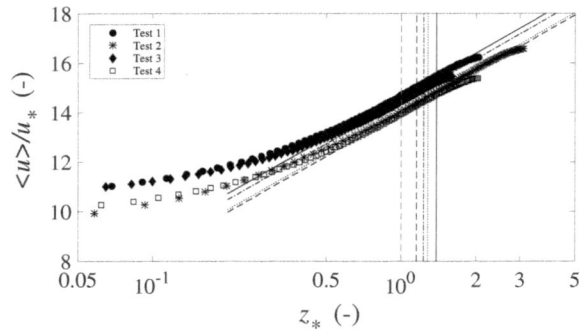

Figure 5. Double-averaged longitudinal velocity profiles and regression lines for Scenario 1 ($\kappa = 0.405$), where $z_* = (z - \Delta)/(k_s - \Delta)$. The red vertical line represents the lower bound of the linear reach for all tests, whereas the black vertical lines define the upper bound determined for each test. Test 1 is identified by a solid line, Test 2 by a dash-dot line, Test 3 by a dashed line, while Test 4 by a dotted line.

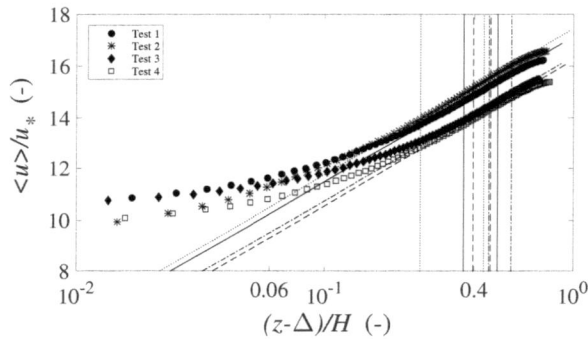

Figure 6. Double-averaged longitudinal velocity profiles and regression lines for the computation of the scale of the roughness elements k_s, for Scenario 2 ($\kappa = 0.405$, $B = 8.5$). The bounds of the regression lines are marked with the solid line (Test 1), dash-dot line (Test 2), dashed line (Test 3), and dotted line (Test 4).

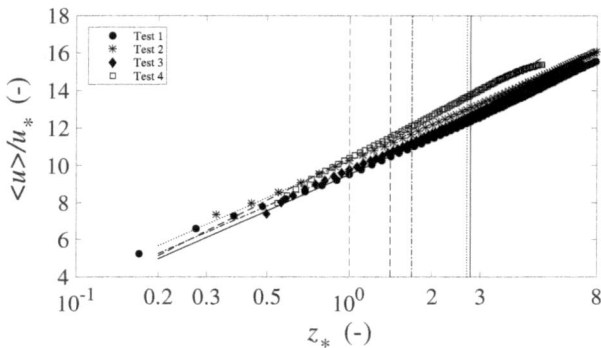

Figure 7. Double-averaged longitudinal velocity profiles and regression lines for Scenario 3, where $z_* = (z - \Delta)/(k_s - \Delta)$. Vertical line specifications are as in Figure 5. The bounds of the regression lines are marked with solid line (Test 1), dash-dot line (Test 2), dashed line (Test 3), and dotted line (Test 4).

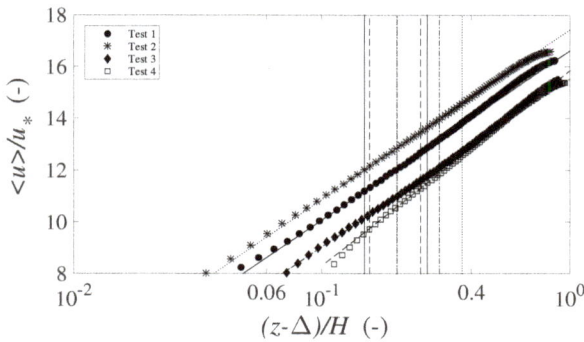

Figure 8. Double-averaged longitudinal velocity profiles and regression lines for the computation of the scale of the roughness elements k_s, for Scenario 4 ($\kappa < 0.405$, $B = 8.5$). The bounds of the regression lines are marked with solid line (Test 1), dash-dot line (Test 2), dashed line (Test 3), and dotted line (Test 4).

Without discussing the nature of κ, the present data show that the adoption of a flow dependent von Kármán parameter, in this case smaller than 0.405 (scenario 3 and 4), is compatible with a log-law layer as an overlapping of inner and outer flow regions. On the contrary, adopting $\kappa = 0.405$ would imply a log-law layer well in the outer region, which is physically difficult to explain.

It should be noticed that the values of the flow dependent κ (scenarios 3 and 4) do not seem, however, correlated with the values of the bedload transport rate, as illustrated in Tables 5 and 6. Irrespectively of the value of the bedload discharge, the von Kármán parameter consistently ranges between 0.305 and 0.355. The scale of the roughness elements k_s increases with the increase of bedload transport rates, both in scenarios 3 and 4, with higher values of k_s found in the latter scenario. The interpretation of k_s as a roughness scale allows for a conjecture about the role of moving bedload particles: since the bed does not suffer strong morphological changes induced by bedload transport, more particles moving should represent a larger work performed by the flow, thus increasing flow resistance, as proposed by Owen [29]. The displacement height Δ, on the contrary, appears to decrease with increasing solid discharge.

3.4. Discussion of Bed Roughness

The results concerning bed roughness are compared with those shown in Ferreira et al. [2], where the velocity profiles of 17 subcritical and nearly uniform flow laboratorial tests are discussed. The channel beds in these two cases are substantially different: while the current study adopted a simple lattice-arranged granular bed with no relevant morphological features even at moderate bedload discharges, the bed surface of Ferreira et al. [2] exhibits a complex micro-topography, with clusters around larger particles, in the case of the armoured beds, and low amplitude bedload sheets, in the case of the sand-gravel mixture at high values of the Shields parameter. Porosity is larger in the current study and tortuosity is expected to be smaller. However, the significant diameters that traditionally determine the value of the scale of the roughness elements—the d_{90} or d_{84}—are, in the tests of Ferreira et al. [2], approximately the same as the diameter of the beads used in the current study. The bedload discharge ranges are equally similar. The simple nature of the bed used in this study allows for an indirect appraisal of the role of the bed micro-topography in determining the geometric scale of the roughness elements and, thus, bottom friction.

The comparison was undertaken for the case where the von Kármán parameter is assumed non-universal (a fitting parameter) and the geometric roughness scale k_s and B are subjected to a best fit procedure (scenario 3).

Figure 9 shows the ratio $\frac{z_0}{d_{50}}$ as a function of the non-dimensional bedload discharge Φ for all 17 tests discussed in Ferreira et al. [2] and for the present experiments.

One interpretation of the results shown in Figure 9 is that bed organization plays a key role in the relationship between roughness height z_0 and d_{50}. In a high complexity system, as the case of the bi-modal mixture of sand and gravel subjected to the armoring process (test of type D described in [2] and represented by black open diamonds), large values of roughness height normalized by the sediment diameter are achieved when bedload transport is incipient (Φ slightly greater than zero). The armoring process, in fact, due to the presence of larger sediments hiding smaller particles from being eroded, produces the maximum topographic diversity in the bed.

In the case of the bed composed by gravel and sand-gravel mixtures, $\frac{z_0}{d_{50}}$ is relatively high at very low sediment transport rates (around 0.08) when compared with the values obtained for the bed composed of glass beads. This shows that bedload movement in the natural bed generates some kind of bed complexity, whereas the simple lattice-arranged does not generate complexity even under bedload conditions. In fact, the ratio z_0/d_{50} in case of a bed matrix made by monosized glass particles presents lower values (between 0.02 and 0.05) and just slightly increase with bedload transport.

The bed micro-topography therefore affects the roughness height and hence the remaining log-law parameters k_s, κ and B. A relevant result is therefore that the roughness height is not correlated with the sediment diameter but depends on bed organization: the more complex the bed topography is, the higher roughness height would be. The large sediment diversity achieved in natural beds responsible for high roughness values can not be obtained with monosized artificial sediments, characterized by a different apparent surface porosity and higher interaction between free-fluid and porous media flows.

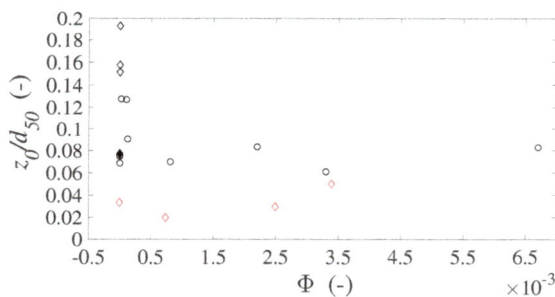

Figure 9. Variation of the roughness height normalized by the sediment diameter d, as a function of the non-dimensional bedload discharge, for scenario 3. Glass particles data are represented by red open diamonds. Data treated by Ferreira et al. [2] of type E are represented by black filled diamonds, type D by black open diamonds and type T by open circles.

4. Conclusions

The experimental analysis reported in this work allowed for testing different definitions of the parameters of the log-law in the specific case of rough mobile bed composed of monosized artificial sediments. Four experimental tests were carried out to assess the effect of bed mobility and increasing bedload transport rates on the wall-normal profile of the longitudinal flow velocity.

The study reveals that a well-fitting log-law profile within the overlapping between outer and inner region ($z/h \leq 0.2$) is achieved assuming a flow-dependent von Kármán parameter κ, together with geometric roughness scale k_s and parameter B subjected to a best fit procedure (scenario 3) or imposing $B = 8.5$ and computing k_s from a roughness function (scenario 4). No substantial dependence of von Kármán parameter on bedload discharge rates was encountered and κ was found ranging between 0.305 and 0.355, depending on the test, contrary to the wider range of κ reported in the literature for beds composed of natural sediment.

The zero of the logarithmic profiles was found very near the sediment crests and slightly increasing with bedload transport (scenario 3). Scenario 4 presented small values of k_s as well, although higher than those achieved in scenario 3.

A comparison between the variation of the roughness height z_0 normalized by the particle diameter as a function of the bedload discharge, obtained in our simple bed configuration and shown in Ferreira et al. [2] for natural beds composed of gravel-sand mixtures, emphasised that the roughness height increases with bed diversity, typical of natural river beds.

The roughness scale k_s is lower in the present tests and higher in the tests with natural sediment, due to the simple configuration of the present bed which prevents the formation of bedload sheets or complex micro-topographical structures.

In the case of glass spheres, the roughness scale increases with bedload transport—it should express the increase in work expended by the flow in maintaining these particles in motion.

The hypothesis that complexity associated with bed micro-topography is more relevant to influence the mean flow than size of the larger elements or bed mobility can thus be put forward and verified.

Author Contributions: Conceptualization, R.M.L.F. and F.A.; methodology, F.A., R.M.L.F. and A.M.R.; formal analysis, F.A., A.M.R. and R.M.L.F.; investigation, F.A., A.M.R. and R.M.L.F.; writing-original draft preparation, F.A.; writing-review and editing, R.M.L.F. and A.M.R., F.A.; visualization, F.A. and A.M.R.; project administration, R.M.L.F.; funding acquisition, R.M.L.F.

Funding: This work was partially funded by Project SEDITRANS funded by the European Commission under the 7th Framework Programme and FEDER, program COMPETE, and national funds through the Portuguese Foundation for Science and Technology (FCT) project MORPHEUS-PTDC/ECM-HID/6387/2014.

Acknowledgments: The authors would like to thank the contribution of Pedro Sanches, Duarte Carona, Cristiana Iannelli, Elena Gatto and João Pedro Caetano in data collection.

Conflicts of Interest: The authors declare no conflict of interest.

References

1. Townsend, A. *The Structure of Turbulent Shear Flow*; Cambridge University Press: Cambridge, UK, 1976.
2. Ferreira, R.M.; Franca, M.J.; Leal, J.G.; Cardoso, A.H. Flow over rough mobile beds: Friction factor and vertical distribution of the longitudinal mean velocity. *Water Resour. Res.* **2012**, *48*, W05529. [CrossRef]
3. Bear, J. *Dynamics of Fluids in Porous Media*, 1st ed.; American Elsevier Publishing Company: New York, NY, USA, 1972.
4. Gaudio, R.; Miglio, A.; Dey, S. Non-universality of von Kármán's κ in fluvial streams. *J. Hydraul. Res.* **2010**, *48*, 658–663. [CrossRef]
5. Nikora, V.; Goring, D. Flow turbulence over fixed and weakly mobile gravel beds. *J. Hydraul. Eng.* **2000**, *126*, 679–690. [CrossRef]
6. Gaudio, R.; Miglio, A.; Calomino, F. Friction factor and von Kármán's κ in open channels with bed-load. *J. Hydraul. Res.* **2011**, *49*, 239–247. [CrossRef]
7. Koll, K. Parameterisation of the vertical velocity profile in the wall region over rough surfaces. In Proceedings of the River Flow 2006, Lisbon, Portugal, 6–8 September 2006; pp. 163–172.
8. Hanmaiahgari, P.R.; Roussinova, V.; Balachandar, R. Turbulence characteristics of flow in an open channel with temporally varying mobile bedforms. *J. Hydrol. Hydromech.* **2017**, *65*, 35–48. [CrossRef]
9. Ferreira, R.M. The von Kármán constant for flows over rough mobile beds. Lessons learned from dimensional analysis and similarity. *Adv. Water Resour.* **2015**, *81*, 19–32. [CrossRef]
10. Pope, S.B. *Turbulent Flows*; Cambridge University Press: Cambridge, UK, 2000.
11. Rodi, W. Turbulence modeling and simulation in hydraulics: A historical review. *J. Hydraul. Eng.* **2017**, *143*, 03117001. [CrossRef]
12. Kalitzin, G.; Medic, G.; Iaccarino, G.; Durbin, P. Near-wall behavior of RANS turbulence models and implications for wall functions. *J. Comput. Phys.* **2005**, *204*, 265–291. [CrossRef]
13. Fröhlich, J.; Von Terzi, D. Hybrid LES/RANS methods for the simulation of turbulent flows. *Prog. Aerosp. Sci.* **2008**, *44*, 349–377. [CrossRef]

14. Piomelli, U. Large eddy simulations in 2030 and beyond. *Philos. Trans. R. Soc. A Math. Phys. Eng. Sci.* **2014**, *372*, 20130320. [CrossRef]
15. Cannata, G.; Petrelli, C.; Barsi, L.; Camilli, F.; Gallerano, F. 3D free surface flow simulations based on the integral form of the equations of motion. *WSEAS Trans. Fluid Mech.* **2017**, *12*, 166–175.
16. Cea, L.; Puertas, J.; Vázquez-Cendón, M.E. Depth averaged modelling of turbulent shallow water flow with wet-dry fronts. *Arch. Comput. Methods Eng.* **2007**, *14*, 303–341. [CrossRef]
17. Williams, H.E.; Briganti, R.; Pullen, T. The role of offshore boundary conditions in the uncertainty of numerical prediction of wave overtopping using non-linear shallow water equations. *Coast. Eng.* **2014**, *89*, 30–44. [CrossRef]
18. Gallerano, F.; Cannata, G.; De Gaudenzi, O.; Scarpone, S. Modeling bed evolution using weakly coupled phase-resolving wave model and wave-averaged sediment transport model. *Coast. Eng. J.* **2016**, *58*, 1650011. [CrossRef]
19. Nezu, I.; Nakagawa, H.; Jirka, G.H. Turbulence in open-channel flows. *J. Hydraul. Eng.* **1994**, *120*, 1235–1237. [CrossRef]
20. Soulsby, R.; Dyer, K. The form of the near-bed velocity profile in a tidally accelerating flow. *J. Geophys. Res. Oceans* **1981**, *86*, 8067–8074. [CrossRef]
21. Stapleton, K.; Huntley, D. Seabed stress determinations using the inertial dissipation method and the turbulent kinetic energy method. *Earth Surf. Processes Landf.* **1995**, *20*, 807–815. [CrossRef]
22. Mendes, L.; Antico, F.; Sanches, P.; Alegria, F.; Aleixo, R.; Ferreira, R.M. A particle counting system for calculation of bedload fluxes. *Meas. Sci. Technol.* **2016**, *27*, 125305. [CrossRef]
23. Ferreira, R.M.; Aleixo, R. *Experimental Hydraulics: Methods, Instrumentation, Data Processing and Management: Vol. II: Instrumentation and Measurement Techniques*; CRC Press, Taylor and Francis Group: London, UK, 2017; Chapter 3, pp. 35–209.
24. Antico, F. Laboratory Investigation on the Motion of Sediment Particles in Cohesionless Mobile Beds under Turbulent Flows. Ph.D. Thesis, Instituto Superior Técnico, Universidade de Lisboa, Lisbon, Portugal, 2018.
25. Nikora, V.; Ballio, F.; Coleman, S.; Pokrajac, D. Spatially averaged flows over mobile rough beds: Definitions, averaging theorems, and conservation equations. *J. Hydraul. Eng.* **2013**, *139*, 803–811. [CrossRef]
26. Ferreira, R.M.; Ferreira, L.M.; Ricardo, A.M.; Franca, M.J. Impacts of sand transport on flow variables and dissolved oxygen in gravel-bed streams suitable for salmonid spawning. *River Res. Appl.* **2010**, *26*, 414–438. [CrossRef]
27. Ferreira, R.M.; Franca, M.; Leal, J. Flow resistance in open-channel flows with mobile hydraulically rough beds. *River Flow* **2008**, *1*, 385–394.
28. Schlichting, H. *Boundary-Layer Theory*, 6th ed.; McGraw Hill: New York, NY, USA, 1968.
29. Owen, P.R. Saltation of uniform grains in air. *J. Fluid Mech.* **1964**, *20*, 225–242. [CrossRef]

water

MDPI

Article

Bed-Load Transport in Two Different-Sized Mountain Catchments: Mlynne and Lososina Streams, Polish Carpathians

Artur Radecki-Pawlik [1], Piotr Kuboń [1,*], Bartosz Radecki-Pawlik [1] and Karol Plesiński [2]

[1]   Institute of Structural Mechanics, Cracow University of Technology, 31-155 Kraków, Poland;
      rmradeck@cyf-kr.edu.pl (A.R.-P.); bradecki-pawlik@pk.edu.pl (B.R.-P.)
[2]   Faculty of Environmental Engineering and Land Surveying, University of Agriculture in Krakow,
      30-059 Krakow, Poland; karol.plesinski@urk.edu.pl
*   Correspondence: pkubon@pk.edu.pl; Tel.: +48-12-628-23-91

Received: 14 January 2019; Accepted: 30 January 2019; Published: 4 February 2019

Abstract: The prediction and calculation of the volume of gravel and/or sand transported down streams and rivers—called bed-load transport is one of the most difficult things for river engineers and designers because, in addition to field measurements, personnel involved in such activities need to be highly experienced. Bed-load transport treated by many engineers marginally or omitted and often receives only minor consideration from engineers or may be entirely disregarded simply because they do not know how to address the issue—in many cases, this is a fundamental problem in river management tasks such as: flood protection works; river bank protection works against erosion; building bridges and culverts; building water reservoirs and dams; checking dams and any other hydraulic structures. Thus, to share our experience in our paper, bed-load transport was calculated in two river/stream mountain catchments, which are different in terms of the characteristics of the catchment area and the level of river engineering works performed along the stream channel—both are tributaries of the Dunajec River and have similar Carpathian flysh geology. The studies were performed in the Mlyne stream and in the Lososina River in Polish Carpathians. Mlynne is one of the streams in the Gorce Mountains—it is prone to flash flooding events and has caused many problems with floods in the past. It flows partially in the natural river channel and partially in a trained river channel lined with concrete revetments. The stream bed load is accumulated in the reservoir upstream of the check dam. The Lososina River is one of the Polish Carpathian mountainous streams which crosses the south of the Beskid Wyspowy Mountains. It mostly has a gravel bed and it is flashy and experiences frequent flooding spring. At the mouth of the Lososina River, there is one of the largest Polish Carpathian artificial lakes—the Czchow lake. The Lososina mostly transports gravel as the bed load to the Czchow water reservoir where the sediment is deposited. In the early seventies, the Lososina was partly canalised, especially in places where passes inhabited areas. The paper compares the situation of bed-load transport in the Lososina River before and after engineering training works showing how much sediment is transported downstream along the river channel to the Czchow artificial lake. Also compared is the Mlynne bed load transport upstream and downstream from the check dam showing how much sediment might be transported and deposited in the reservoir upstream from the check dam and when one could expect this reservoir to be clogged.

Keywords: mountain stream; Mountain River; check dam; water reservoir; bed-load transport

1. Introduction

Bed-load transport measurements and its calculations in streams and rivers is of upmost importance in many technical, engineering and fluvial-associated activities but it is very difficult.

It demands not only skills and knowledge but also scientists and designers with many years of personal experience in the field. It is not a question of the models used or the applied methodology—it is very often a question of how we 'feel' the river or the stream and how skilfully we can conduct field measurements. Even after careful calculations and field measurements when working with bed-load transport for the Tatra National Park in the Carpathians, we contacted some international colleagues to ensure that our predictions of bed-load transport were correct due to it being the first time that we performed bed-load transport calculations at such a scale. Moreover, we are using their literature [1,2] in this paper to further improve our understanding of sediment transport since it is either treated marginally by many engineers or may even be totally omitted, simply because they do not know how to deal with the issue of bed-load transport interpretation.

Sediment transport is in many cases a fundamental problem in river management tasks such as flood protection works, river bank protection works against erosion, building bridges and culverts and building water reservoirs, dams, check dams and any other hydraulic structures. This is our motivation for producing a paper in the hope that it promotes a better understanding of the bed-load transport phenomena. The highest priority problem in our paper is bed-load transport which might cause the clogging of water reservoirs built further downstream the river. Obviously, some check dams are built to trap sediment in upstream reservoirs but the question remains of whether it is a worthwhile expense time, money and effort to build them for the sake of flood protection or whether it is more appropriate to simply let the sediment move to the main river. The question is especially interesting when we try to compare small and huge water reservoirs in very similar fluvial and geological situations to those we have in our case, since we work in one large catchment of the Dunajec river in the Polish flysch Carpathian mountains; however, we are still considering two tributaries to the Dunajec which differ in the size of the sub-catchments and in terms of the size of water reservoirs built there.

The question of sediment transport and water reservoirs has arisen so often recently when considering whether it is worthwhile to deliberately breach dams and check dams and/or remove them and fill up water reservoirs with sediment thus rehabilitating rivers and rivers valleys [3–8]. The aim of this paper is to show the difference in bed-load transport in two different rivers which might provide valuable information for river management with regard to how to deal with hydraulic river infrastructure built in their catchments for the future including decisions relating to the removal of dams and water reservoirs.

Just to introduce a reader in the sediment transport phenomena one has to bear in mind that water and sediment in rivers has an enormous impact both on the environment and on people. Rivers very often change their cross sections and longitudinal profiles as a result of the process of sediment transport [9]. By moving, rolling, skipping or sliding downstream along the river channel, the sediment refers to the form of the bed load, which is transported to the river mouth [10–13]. The longitudinal profile is also shaped by the flowing water and sediment.

Generally, one can distinguish between two types of sediments in rivers: the bed load and the suspended load [9,14]. In mountain streams where the streambed consists mostly of gravel and coarse sands, the bed load is reported to constitute in some cases even up to 70 per cent of the total bed load [15]. Mountain stream gravel is very often both legally and illegally mined from riverbeds, which is disastrous for the fluvial state of rivers and for river ecology; furthermore, it causes the destruction of flood protection strategies, river revetments, bridges and all hydraulic structures [16–20]. This situation presents a major problem for all river managers. The bed load of the mountain streams in the Polish Carpathians has been the subject of many scientific studies in which hydraulic structures and river training problems are considered [21–25]. The problem caused by the movement of sediment is especially dangerous when we have low-head hydraulic structures built along streams or rivers [26–32]. Furthermore, it presents problems with river ecohydrology [33–35] and when the water reservoir for flood protection and for water storage is constructed on the river because the sediment trapped in the reservoir tends to fill it up since it is resulting in there being greatly reduced capacity for water [9,21,36,37]. This situation occurs on the Lososina River and is one of the subjects of this paper. To reduce this process (reducing the bed-load transport) many engineering works are undertaken—these are referred to as river-training works.

Some river-training works take the form of constructing check dams and some are simply the installation of river sills aimed at reducing the gradient of the river slope [36–39]. This situation applies to the Mlynne stream, which is also described in this paper. Both conditions of bed-load transport in the presented two cases are sources of valuable information for river managers dealing with sediment problems. The novel aspect of our paper is the presentation of two cases of bed-load transport which are analysed in two different-sized sub catchments of two rivers with artificial water reservoirs, which are both tributaries of a larger river—the Dunajec. The geology of the region is similar. The reader and the potential river manager should be acquainted with such research in order to have a sense of the scale of problem and to remember that only the careful and professional analysis of streams and rivers can assist in making decisions when bed-load transport information is needed. Thus, the conclusions presented in the article may enable them to decide what actions may be required to improve the hydromorphological conditions in the case of similar mountain streams.

2. Study Areas

2.1. The Mlynne Stream

The Mlynne catchment is part of the Western Carpathian Province, the Outer Western Carpathians sub province, the Outer Western Beskids macroregion and the Gorce Mesoregion. The Mlynne stream is left tributary of the Ochotnica stream (right Dunajec river tributary) [40]. On the lower reach (from the check dam to the mouth) the terrain elevation gradient is around 90 m and the average slope is 3.6 per cent. The average slope of the upper reach is 10.2 per cent and the stream valley development V (calculated according to [41]) is 0.364. The orographic index (λ) of 777.48 according to [42] classifies the Mlynne stream as a high-mountain watercourse and the Łochtin stability parameter (f) of 0.913 [43] defines the Mlynne stream riverbed as being vulnerable to erosion.

The whole Mlynne catchment area lies within the Gorce Mountains, built in the most part from Magura Set sedimentary rocks. The rocks of this set cover the largest area in Outer Western Flysch Carpathians and build the Zywiecki Beskid Mountains, the larger part of Medium Beskids, Insular Beskids, Sadecki Beskids and the part of Low Beskids. Sedimentary rocks are classed as so-called flysch here, consisting of alternating layers of sandstones, mud shales, pudding stones, mudstones and siltstones. The flysch is often accompanied by carbonate rocks, such as limestones, marls and dolomites, also found in the Gorce Mountains [40]. In the Mlynne stream valley, ensembles of thick- and thin-shoaled sandstones, shifted by greyish shales, are being exposed. Additionally, abundant rock verges occur in the Mlynne stream riverbed, especially in its upper reach. The mica sandstones and shales of the Magura layer dominate in that section. Physical characteristics of the Młynne stream is presented in Table 1.

In the middle catchment area there are mainly Tertiary (Paleogene) shales and sub-Magura layer sandstones with the lens of Quaternary slide colluviums. Sparse alluvial settlements are also present in the Mlynne stream valley in the lower part of the catchment area [44]. The measurements were performed at km 0.0 (Mlynne outlet to Ochotnica—495.00 m above sea level) to km 7 + 500 (985.00 m above sea level). The Mlynne stream was divided into six measurement sections (Figure 1); in each section, the following test sections were determined: cross section 1-1 at km 0 + 150 (498.00 m above sea level called 'bystrotok'); cross sections 2-2, 3-3, 4-4 at km 2 + 800 (at 85.00 m above sea level in the area of the reservoir behind the dam called 'reservoir'—the sample from section 2-2 was taken from a small, stable sediment deposited just behind the notch of the check dam; the sample from section 3-3 was taken from the river channel in the vicinity of the mainstream stream channel at the check dam water reservoir; sample 4-4 was taken from the reservoir edge, from the inlet to the reservoir); section 5-5 at km 3 + 300 (610.00 m above sea level, the section located in the built-up area in the natural part of the channel called 'school'); section 6-6 in km 4 + 300 (655.00 m above sea level) in a natural state called 'Kotelniki').

Here, we are only presenting bed-load transport calculations for cross sections 2-2 'check-dam reservoir,' cross section 5-5 'school' and 6-6 'Kotelniki'. This approach was chosen because we wanted to show the deposition in cross section 2-2 (in the water reservoir formed upstream of the check dam) and the difference in the sediment transport upstream of this place as we are dealing with natural and partly engineered cross sections upstream (5-5 and 6-6). In cross sections 3-3 and 4-4, we have taken sediment transport samples for grain-size reasons to have the average grain size curve for the whole water reservoir upstream of the check dam. Cross section 1-1 is downstream of the check dam and along the whole section (longitudinal profile from the check dam down to the estuary of the Mlynne) both river banks are lined with concrete and artificial boulders which creates a kind of rapid channel. Thus, we do not present bed-load transport values here (sediment is trapped upstream of this section); however, we performed some hydraulics measurements here that are not presented in this paper but which were helpful in understanding the whole regime of the stream; see Figure 1 for details. Exemplary natural as well as modified cross sections of the Młynne Stream are presented in Figures 2 and 3 respectively.

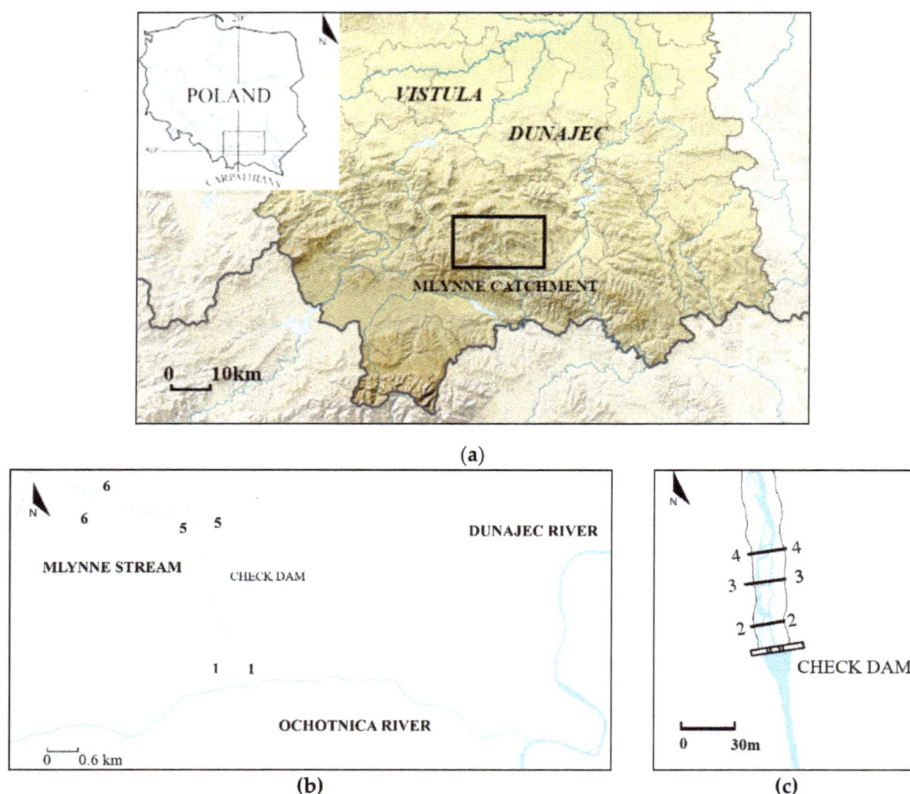

Figure 1. Location of the Mlynne Stream catchment (**a**); the chosen research cross sections and research reach (**b**); check dam cross sections details (**c**).

Table 1. Physical characteristics of investigated sites the Mlynne Stream.

Variables	The Mlynne Stream
precipitation (mm)	850
catchment area (km^2)	7.3
max. catchment area altitude (m a.s.l.)	985.00
min. catchment area altitude (m a.s.l.)	495.00
channel gradient (average within study area) (-)	0.022
max. stream length (km)	7.50
T-year flood $Q_{50\%}$ (m$^3 \cdot$s^{-1})	7.70
T-year flood $Q_{5\%}$ (m$^3 \cdot$s^{-1})	40.4
d_{16} (mm)	7
d_{50} (mm)	39
d_{84} (mm)	94
d_{90} (mm)	102

(**a**)   (**b**)

Figure 2. Mlynne stream—natural cross sections. (**a**) upper part not engineered; (**b**) source part.

(**a**)   (**b**)

Figure 3. *Cont.*

(c)   (d)

Figure 3. Mlynne stream—cross sections modified with check dam and water reservoir upstream of the check dam. (**a**) check-dam; (**b**) check-dam reservoir with coarse sediment; (**c**) down stream of a check-dam; (**d**) fine sediment in a check-dam reservoir.

2.2. The Lososina River

The Lososina River in the Polish part of the Carpathian Mountains (Figure 4) is situated in the Carpathian flysh. The stream is flashy and experiences frequent bed-load movement. Its streambed consists mostly of sandstone and mudstone bed-load pebbles and cobbles forming a framework, the interstices of which are filled by a matrix of finer sediment. Exemplary cross sections of the Lososina River are presented in Figure 5.

(a)

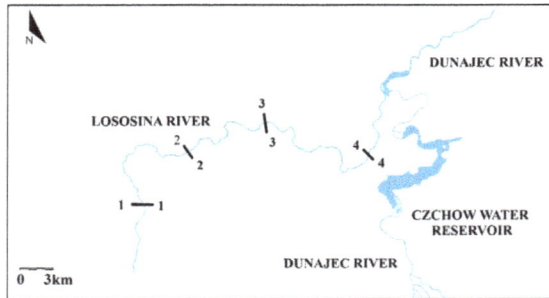

(b)

Figure 4. Catchment study region of the Lososina River (**a**) with a detailed sketch of the research reach (**b**).

Figure 5. The Lososina River—different cross sections along the river reach. (**a**) source part in Dobra; (**b**) engineered part in Podplomien; (**c**) engineered part in Tymbark; (**d**) river training works in Laskowa; (**e**) natural part in Lososina; (**f**) water reservoir Czchow.

The suspended sediment load is small but contributes to channel morphology—this was taken into consideration during sediment calculations. Many gravel river bed-forms, such as point and middle bars, can be observed within the investigated Lososina River. Most gravel-bed forms can be observed on the riverbanks and within the river channel. After 1975, many river-training works were performed along the Lososina channel to prevent bank erosion and to reduce the channel slope. The river cross sections were trained by building drop-hydraulic structures (for the purpose of slope reduction) and by constructing gabions (stone-baskets along the banks to prevent bank erosion). Those works were aimed at reducing the bed-load transport along the Lososina and stopping its degradation, since the Czchów river reservoir was constructed at the river mouth. The basic hydrological characteristics and grainsize characteristics of the river are presented in Tables 2 and 3. All numbers refer to the entire river channel between cross sections 1-1 and 4-4.

Table 2. Physical characteristics of investigated sites—the Lososina River.

Variables	The Lososina River
precipitation (mm)	896
catchment area (km^2)	410
max. catchment area altitude (m a.s.l.)	760.00
min. catchment area altitude (m a.s.l.)	241.00
channel gradient (average within study area) (-)	0.011
max. stream length L (km)	49.00
discharge $Q_{50\%}$ (m^3 s^{-1})/flood $Q_{50\%}$	48.63
discharge $Q_{5\%}$ (m^3 s^{-1})/flood $Q_{3\%}$	196.41

Table 3. Characteristic grain size for the Lososina River before and after river-training works.

Sampled Cross Section	Before River-Training Works				After River-Training Works			
	Sediment Diameter (mm)				Sediment Diameter (mm)			
	d_{16}	d_{50}	d_{84}	d_{90}	d_{16}	d_{50}	d_{84}	d_{90}
1-1	7	28	83	88	7	30	85	90
2-2	10	30	70	76	6	22	65	70
3-3	12	40	90	95	10	35	88	90
4-4	10	30	58	67	11	22	50	65

3. Methods

Bed-load transport for both the Lososina River and the Mlynne stream was calculated using Meyer-Petter Muller [45] formula:

$$q_i = \left[\frac{\rho_w \cdot g \cdot h \cdot I - f_i \cdot g \cdot \Delta\rho \cdot d_i}{0.25 \cdot \rho_w^{\frac{1}{3}}} \right]^{1.5} p_i \times b \ (\text{kg s}^{-1} \text{ m}^{-1}) \tag{1}$$

where q_i—unit bed-load transport [N s^{-1}]; ρ_w, ρ_r—water and sediment density (kg m^{-3}); g—acceleration (m s^{-2}); h—water depth (m); I—slope (-); f_i—shields shear stress value (-); $\Delta\rho = \rho_r - \rho_w$ (kg m^{-3}); d_i—sediment size (mm); p_i percentage of the sediment fraction within the sediment probe, b—active channel width (m).

According to Michalik [9], the dimensionless shear stress parameter for Polish mountain streams was identified using radioisotope methods and is 0.033—this is taken for all calculations in this paper, although one has to remember that the original Meyer-Petter Muller [45] dimensionless shear stress was 0.047. Field survey and slopes measurements were performed with a TOPCON AT-G7 survey professional level device. The bedload transport value was calculated using the SandCalc 1 software application [46].

In order to understand the hydrological situation of the stream and rivers, some characteristic discharges are usually calculated—this sheds light on how considerable flood events are currently dealt with in river channels. One method used for this purpose is the calculation of T-year floods. As defined by Frost and Clark [47], the T-year flood is a discharge likely to be exceeded once in T-years on average. For example, the 100-year flood is also referred to as the 1% flood, since its annual exceedance probability is 1%. T-year flood values Q for the Mlynne stream estuary and the Lososina stream estuary were calculated using the Punzet method [48,49]—Tables 1 and 2.

For sediment analysis, the grain size curves were performed on the basis of a classical sieving survey [10]. All the basic granulometric parameters calculated for each of the researched cross sections are presented in Tables 1 and 3 (for the Mlynne and the Lososina streams, respectively).

4. Results and Discussion

4.1. The Młynne Stream

For the sake of clarity, firstly the results of measurements and calculations referring to the bed-load transport are presented in tables and graphs. Table 4 presents bed-load transport results in cross sections 2-2 'check-dam reservoir', cross section 5-5 'school' as well as for 6-6 'Kotelniki'.

Table 4. Unit bed-load transport measured in sampling cross sections: 2-2 'check-dam reservoir'; 5-5 'school' and 6-6 'Kotelniki'.

Unit Bedload Transport (kg s−1 m−1)						
Water Depth h (m)	Sampling Cross Section 2-2	Depth h (m)	Sampling Cross Section 5-5	Depth h(m)	Sampling Cross Section 6-6	
0.10	0.643	0.10	no bed-load transport	0.10	no bed-load transport	
0.30	1.338	0.30	no bed-load transport	0.30	no bed-load transport	
0.50	1.880	0.50	20.7308×10^{-5}	0.50	7.4031×10^{-5}	
0.70	2.303	0.70	48.1816×10^{-5}	0.70	2.6798×10^{-5}	
0.90	2.028	0.90	161.1152×10^{-5}	0.90	54.4528×10^{-5}	
1.20	2.878	1.20	165.7039×10^{-5}	1.20	101.7977×10^{-5}	
Total	53.5249×10^{-5}	Total	395.7315×10^{-5}	Total	166.3035×10^{-5}	

The bed load transport in all the analysed cross sections of the Młynne stream is small. Its range is 53.5249×10^{-5}–166.3035×10^{-5} (kg s^{-1} m^{-1}), while in other Polish Carpathian streams investigated by Michalik [22], the measured sediment transport where the radioisotope methods were used was, in the Wisłoka Stream: 183.5489×10^{-5}–1488.7857×10^{-5} (kg s^{-1} m^{-1}), in the Raba Stream: 611.8297×10^{-5}–7138.0135×10^{-5} (kg s^{-1} m^{-1}) and in the Dunajec River: 3161.1203×10^{-5}–3467.0351×10^{-5} (kg s^{-1} m^{-1}). North American streams with a flow regime similar to that of the Młynne, such as East Fork in Wyoming, Snake River in Idaho and Mountain Creek in South California, are characterised by higher transport values of: 1019.7162×10^{-5}–7138.0135×10^{-5} (kg s^{-1} m^{-1}), 1019.7162×10^{-5}–$10197.1621 \times 10^{-5}$ (kg s^{-1} m^{-1}) and 50.9858×10^{-5}–1019.7162×10^{-5} (kg s^{-1} m^{-1}), respectively [11]. Such a small amount of sediment transported might suggest that the necessity of the check dam built in the stream is questionable [36]. It could transpire that rehabilitation works, which could be planned in that catchment take into consideration removing the existing dam with no harm for the sediment budget and in line with stream restoration works advice at present in Polish Carpathians [33–35].

4.2. The Lososina River

Again, all obtained results are presented in tables and in graphs for the benefit of clarity. Tables 5 and 6 show the sediment transport data for Lososina before and after river-training works [50–54]. Table 7 presents changes of the unit bed-load transport results for the Lososina River after regulation. Figure 6 presents the hydrological events before and after river-training works at the Lososina used in the TransCalc computer model to calculate the sediment budget along the Lososina.

As can be observed, despite the decreasing of sediment dimensions within the river-trained cross sections, the shear stress values there also decreased; consequently, the unit bed-load volume decreased as well. The most important parameter here, which determines the value of the decreased shear stresses and the bed load, is reduced slope by river training [37,39]. Here, in the 4-4 cross section, the largest decrease of bed-load transport was identified—this was the main aim of the river-training works. Due to river engineering works, the change of the bed-load transport along the Lososina river in the analysed cross sections was as follows: along cross section 1-1, the unit bed load q was larger before the river training at around 1.8121 kg s^{-1} m^{-1} (aggradations after river-training works), along cross section 2-2 the unit bed-load q was also larger before the river training of about 16.5302 kg s^{-1} m^{-1} (aggradations after river-training works). Along cross section 3-3, the unit bed load q was larger after river training of about 4.5341 kg s^{-1} m^{-1} (degradation of the river bed) and finally, along cross section 4-4, the unit bed load q was larger before the river training of about q = 17.9028 kg s^{-1} m^{-1}. Along the whole river, the unit bed load q was bigger before the river training of about q = 13.3687 kg s^{-1} m^{-1}, in other words, the river training reduced the bed-load transport of that value. The river training works performed along the Lososina river changed the bed-load transport conditions along its entire length [29,38,55,56].

Table 5. Unit bed-load transport at the Lososina River—prior to the training works.

Water Depth h (m)	Sampling Cross Section 1-1 d_{50} = 28 (mm)		Sampling Cross Section 2-2 d_{50} = 28 (mm)		Sampling Cross Section 3-3 d_{50} = 28 (mm)		Sampling Cross Section 4-4 d_{50} = 28 (mm)	
	Shear Stress τ (N m^{-2})	Transport (kg s^{-1} m^{-1})	Shear Stress τ (N m^{-2})	Transport (kg s^{-1} m^{-1})	Shear Stress τ (N m^{-2})	Transport (kg s^{-1} m^{-1})	Shear Stress τ (N m^{-2})	Transport (kg s^{-1} m^{-1})
0.6	no bed-load transport							
0.7	21.39	0.0112						
0.8	23.24	0.1439						
0.9	28.07	0.7892	no bed-load transport		no bed-load transport		no bed-load transport	
1.0	29.28	0.9990						
1.1	32.38	1.6059						
1.2	34.67	2.1094						
1.3	43.76	4.5236	25.78	0.2520				
1.4			28.24	0.5793	33.47	0.2764	22.65	0.0037
1.5			32.28	1.2797	39.36	1.2034	27.42	0.4600
1.6	max. depth in		35.60	1.9796	49.38	3.5642	32.57	1.3383
1.7	cross section		41.33	3.4106	52.34	4.4070	38.09	2.5705
1.8	1. 3 (m)		43.98	4.1499	55.38	5.3318	43.98	4.1499
1.9			50.23	6.0818	58.49	6.3414	50.23	6.0818
2.0			53.09	7.0490	61.69	7.4332	53.09	7.0490

Table 6. Unit bed-load transport at the Lososina River—after the training works.

Water Depth h (m)	Sampling Cross Section 1-1 d_{50} = 28 (mm)		Sampling Cross Section 2-2 d_{50} = 28 (mm)		Sampling Cross Section 3-3 d_{50} = 28 (mm)		Sampling Cross Section 4-4 d_{50} = 28 (mm)	
	Shear Stress τ (N m^{-2})	Transport (kg s^{-1} m^{-1})	Shear Stress τ (N m^{-2})	Transport (kg s^{-1} m^{-1})	Shear Stress τ (N m^{-2})	Transport (kg s^{-1} m^{-1})	Shear Stress τ (N m^{-2})	Transport (kg s^{-1} m^{-1})
0.8	no bed-load transport				no bed-load transport			
0.9	22.85	0.0103						
1.0	27.78	0.5177			28.43	0.1389	no bed-load transport	
1.1	33.03	1.4284			30.68	0.3956		
1.2	35.36	1.9266	no bed-load transport		31.77	0.5475		
1.3	44.63	4.3380			32.28	0.6246	17.36	0.0352
1.4					33.46	0.8136	18.27	0.1010
1.5					44.77	3.3258	19.21	0.1892
1.6	max. depth in		17.80	0.0639	47.54	4.0969	19.95	0.2708
1.7	cross section		24.21	0.8960	50.38	4.9422	20.94	0.3935
1.8	1. 3 (m)		26.26	1.2720	53.31	5.8671	21.95	0.5340
1.9			28.38	1.7092	56.30	6.8662	25.29	1.0897
2.0			30.57	2.1995	59.38	7.9420	25.72	1.1708

Table 7. Budget of the unit bed-load transport—the Lososina River.

Unit Bed—Load Transport (kg s^{-1} m^{-1})							
Depth h (m)	Sampling Cross Section 1-1	Depth h (m)	Sampling Cross Section 2-2	Depth h (m)	Sampling Cross Section 3-3	Depth h (m)	Sampling Cross Section 4-4
0.9	0.7788	1.6	1.9157	1.5	−2.1200	1.4	−0.0973
1.0	0.4873	1.7	2.5146	1.6	−0.3100	1.5	0.2708
1.1	0.1776	1.8	2.8779	1.7	−0.5353	1.6	1.0674
1.2	0.1829	1.9	4.3725	1.8	−0.5353	1.7	2.1770
1.3	0.1855	2.0	4.8495	1.9	−0.5247	1.8	3.6146
-	-	-	-	2.0	−0.5088	1.9	4.9921
-	-	-	-	-	-	2.0	5.8782
Total	1.8121	Total	16.5302	Total	−4.5341	Total	17.9028

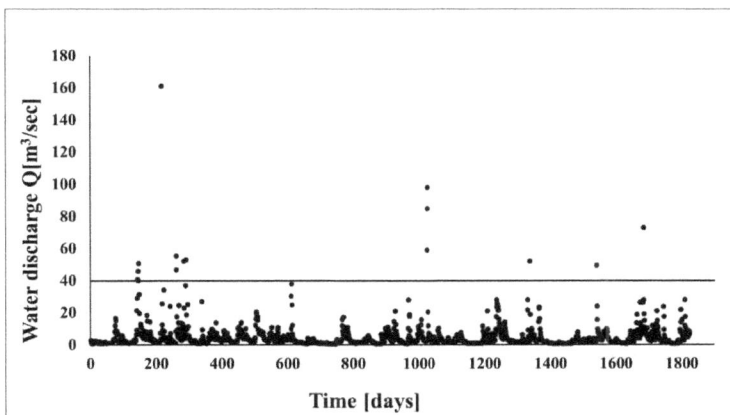

(a)

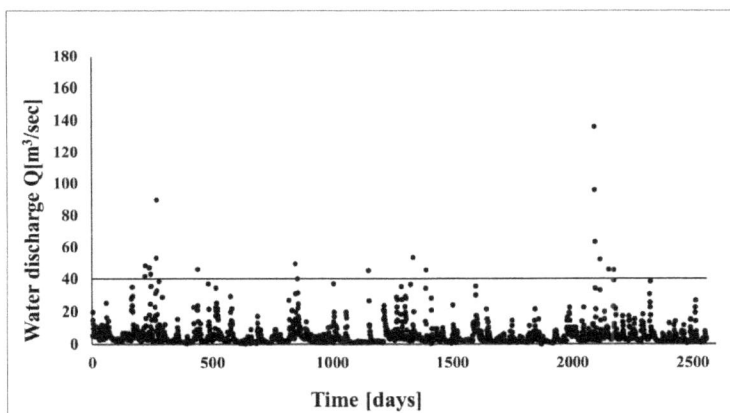

(b)

Figure 6. Water discharge levels for the Lososina river with the threshold line (the beginning of motion for the sediment) above which the bed-load transport was calculated before (**a**) and after (**b**) the river-training works performed in 1975. The horizontal lines are showing the discharge value above which bedload transport occurs.

For both analysed catchments—looking at numbers—one is supposed to find some similarities or dissimilarities and find the problems, which occur in that case in river channels. As we might observed in the Mlynne stream, the bed load transport is low. As we also observed in the field, people from the surrounding terrain who live within the Mlynne catchment remove sediment from the water reservoir upstream of the Mlynne dam after floods and use it for building purposes as well as for strengthening their private land with gravel and clay, depend if they do it on roads or on agriculture fields. Basically, they remove the sediment and in this way, they enlarge the water reservoir volume so it is ready for the sediment from the next flood. This raises the question of whether removing the check dam might be possible because of the low bed-load transport rate. The answer is complicated since downstream of the check dam, there is a road parallel to the engineered stream and if the check dam is removed, sooner or later the road would be covered with sediment. However, one might consider enlarging the river bed downstream of the check dam and when removing it along that area, the Mlynne might to start to be a braided channel again. From a river-management point of view, such a situation is possible but difficult to force upon the local community.

The strengths of the presented approach is that we are gaining knowledge and it may lead us to the identification of easier tools for river management in the future. The weaknesses of the approach are that, although we have increased our understanding of the problems, we cannot force rivers and streams to be braided again because of the local community living in the villages; furthermore, it is difficult to propose the removal of dams as flood protection is traditionally associated with solid hydraulic structures such as dams. Thus, as far as river management is concerned, we need to wait for a really large flood event which could destroy the Mlynne check dam and we might never rebuild it, safe in the knowledge that sediment transport here is low.

In terms of the Lososina, we surprisingly see a good situation of sediment budget after the wisely designed river engineering works. The sediment transported to the Lososina water reservoir is relatively low—this gives the likelihood of the longevity of this water reservoir. In both cases, our study was performed with the specific geology of the Carpathians so one has to bear in mind the limitation of this study to Flysh Mountains. However, the general conclusions are useful for all river management works and might be used for numerical models to improve the performance estimating bed load transport.

5. Conclusions

For both analysed the catchments, the final conclusions are as follows:

1.   During the floods, the Mlynne Stream transports, in comparison with other mountain streams, there is a lower bed load (53.5249×10^{-5}–166.303×10^{-5} kg s^{-1} m^{-1}). This might be connected with the catchment area as well as with the channel slope.

2.   Since the bed-load transport for the Mlynne is marginal, the need for the existence of the check dam is questionable and in the future it could be possible to deconstruct and remove it. Such a practice would be in line with the rehabilitation works started on the Carpathian streams which are already leading to the reconstruction of braided gravel mountain streams.

3.   Because of reducing the river slope of the longitudinal profile of the Lososina when it was river trained, the shear stress values decreased. As a consequence of this, the unit bed-load volume decreased. This indicates the importance of the slope of a river channel for sediment movement when managing rivers. It might be reached by, for example, a series of hydraulic structures across the river channel.

4.   The Lososina river training reduced the bed-load transport by a value of q = 13.3687 kg s^{-1} m^{-1}. In terms of the water reservoir and its clogging, this is useful information in terms of river management practices because the reservoir is a source of drinking water for the region.

5.   The study was performed to assist river and mountain stream managers and urban-village planners to understand how important it is to include bed-load transport in designing calculations when dealing with any river channel problems. The next step of such research could be an analysis of the hydrological situation after removing the check dam from the stream and/or introducing a new philosophy of river rehabilitation works to the region where the sediment is low whilst at the same time, giving due consideration to flood protection aims. In all cases, knowledge of the sediment budget is eternally helpful.

Author Contributions: Conceptualization, A.R.-P., B.R.-P., K.P. and P.K.; Methodology, A.R.-P., B.R.-P. and P.K.; Investigation, A.R.-P., B.R.-P. and K.P.; Resources, A.R.-P. and B.R.-P.; Writing-Original Draft Preparation, A.R.-P., B.R.-P., P.K. and K.P.; Writing-Review & Editing, A.R.-P. and P.K.; Visualization, A.R.-P. and P.K.; Supervision, A.R.-P.

Funding: This research was financed by the Ministry of Science and Higher Education of the Republic of Poland: 1. Cracow University of Technology, Faculty of Civil Engineering: L4/106/2018/DS, L4/107/2018/DS and L4/585/2018/DS-M.

Conflicts of Interest: The authors declare no conflict of interest.

References

1. Galia, T.; Hradecký, J. Estimation of bedload transport in headwater streams using a numerical model (Moravskoslezké Beskydy Mts, Czech Republic). *Acta Univ. Carol. Geogr.* **2014**, *49*, 21–31. [CrossRef]
2. Church, M.; Hassan, M.A. Size and distance of travel of unconstrained clasts on a streambed. *Water Resour. Res.* **1992**, *28*, 299–303. [CrossRef]
3. Bednarek, A.T. Undamming rivers: A review of the ecological impacts of dam removal. *Environ. Manag.* **2001**, *27*, 803–814. [CrossRef]
4. Stanley, E.H.; Doyle, M.W. Trading off: The ecological effects of dam removal. *Front. Ecol. Environ.* **2003**, *1*, 15–22. [CrossRef]
5. Lane, S.N. Acting, predicting and intervening in a sociohydrological world. *Hydrol. Earth Syst. Sci.* **2014**, *18*, 927–952. [CrossRef]
6. Fox, C.A.; Magilligan, F.J.; Sneddon, C.S. "You kill the dam, you are killing a part of me": Dam removal and the environmental politics of river restoration. *Geoforum* **2016**, *70*, 93–104. [CrossRef]
7. Magilligan, F.J.; Graber, B.E.; Nislow, K.H.; Chipman, J.W.; Sneddon, C.S.; Fox, C.A. River restoration by dam removal: Enhancing connectivity at watershed scales. *Elem. Sci. Anthr.* **2016**, *4*. [CrossRef]
8. Bellmore, R.J.; Duda, J.J.; Craig, L.S.; Greene, S.L.; Torgersen, C.E.; Collins, M.J.; Vittum, K. Status and trends of dam removal research in United States. *Wirel. Water* **2017**, *4*. [CrossRef]
9. Yang, C.T. *Sediment Transport: Theory and Practice*; McGraw-Hill: New York, NY, USA, 1996; ISBN 10 0079122655.
10. Church, M.A.; McLean, D.G.; Wolcot, J.F. River Bed Gravels: Sampling and Analysis. In *Sediment Transport in Gravel-Bed Rivers*; Throne, C.R., Bathurst, J.C., Hey, R.D., Eds.; John Wiley and Sons: London, UK, 1987; pp. 43–87.
11. Gomez, B.; Church, M. *A Catalogue of Equilibrium Bedload Transport Data for Coarse Sand and Gravel-Bed Channels*; University of British Columbia: Vancouver, BC, Canada, 1988.
12. Carling, P.A. Bedload transport in two gravel-bedded streams. *Earth Surf. Process. Landf.* **1987**, *14*, 27–39. [CrossRef]
13. Mrokowska, M.M.; Rowiński, P.M.; Książek, L.; Strużyński, A.; Wyrębek, M.; Radecki-Pawlik, A. Laboratory studies on bedload transport under unsteady flow conditions. *J. Hydrol. Hydromech.* **2018**, *66*, 23–31. [CrossRef]
14. Dey, S. *Fluvial Hydrodynamics: Hydrodynamic and Sediment Transport Phenomena*; Springer: Berlin, Germany, 2014; ISBN 978-3-642-19062-9.
15. Selby, M.J. *Earth's Changing Surface: An Introduction to Geomorphology*; Oxford University Press: London, UK, 1985; ISBN 10 9780198232513.
16. Radecki-Pawlik, A. Pobór żwiru i otoczaków z dna potoków górskich (in Polish). *Aura—Ochr. Środowiska* **2002**, *3*, 17–19.
17. Wyżga, B.; Hajdukiewicz, H.; Radecki-Pawlik, A.; Zawiejska, J. Eksploatacja osadów z koryt rzek górskich—skutki środowiskowe i procedury oceny (in Polish). *Gospod. Wodna* **2010**, *6*, 243–249.
18. Lehotský, M.; Frandofer, M.; Novotný, J.; Rusnák, M.; Szmańda, J.B. Geomorphic/Sedimentary Responses of Rivers to Floods: Case Studies from Slovakia. In *Geomorphological Impacts of Extreme Weather. Case Studies from Central and Eastern Europe*; Lóczy, D., Ed.; Springer: Dordrecht, The Netherlands, 2013; pp. 37–52.
19. Lehotský, M.; Kidová, A.; Rusnák, M. Slovensko-anglické názvoslovie morfológie vodných tokov. *Geomorphol. Slovaca Et Bohem.* **2015**, *15*, 1–63. Available online: http://www.asg.sav.sk/gfsb/v0151/GSeB_1_2015.pdf (accessed on 2 February 2019).
20. Kondolf, G.M. Hungry water: Effects of dams and gravel mining on river channel. *Environ. Manag.* **1997**, *21*, 533–551. [CrossRef]
21. Gladki, H. Discussion of "Determination of Sand Roughness for Fixed Beds". *J. Hydraul. Res.* **1975**, *13*, 221–222. [CrossRef]
22. Michalik, A. Bed-Load Transport Investigations in Some Polish Carpathians Rivers. Ph.D. Thesis, University of Agricultural in Krakow, Krakow, Poland, 1990. (In Polish)

23. Radecki-Pawlik, A. Point bars development and sediment structure in the Skawica Creek in Polish Carpathians. International conference on transport and sedimentation of solid particles. *Zesz. Nauk. Ar We Wrocławiu* **2000**, *320*, 113–120.

24. Mikuś, P.; Wyżga, B.; Radecki-Pawlik, A.; Zawiejska, J.; Amirowicz, A.; Oglęcki, P. Environment-friendly reduction of flood risk and infrastructure damage in a mountain river: Case study of the Czarny Dunajec. *Geomorphology* **2015**, *272*, 43–54. [CrossRef]

25. Hajdukiewicz, H.; Wyżga, B.; Amirowicz, A.; Oglęcki, P.; Radecki-Pawlik, A.; Zawiejska, J.; Mikuś, P. Ecological state of a mountain river before and after a large flood: Implications for river status assessment. *Sci. Total Environ.* **2018**, *610–611*, 244–257. [CrossRef]

26. Pagliara, S.; Radecki-Pawlik, A.; Palermo, M.; Plesiński, K. Block ramps in curved rivers: Morphology analysis and prototype data supported design criteria for mild bed slopes. *River Res. Appl.* **2017**, *33*, 427–437. [CrossRef]

27. Pagliara, S.; Radecki-Pawlik, A.; Palermo, N.; Plesiński, K. A Preliminary Study of field scour morphology downstream of block ramps located at river bends. In Proceedings of the 7th IAHR International Symposium on Hydraulic Structures, Aachen, Germany, 15–18 May 2018; Bung, D., Tullis, B., Eds.; Utah State University: Logan, UT, USA, 2018.

28. Sattar, A.M.A.; Plesiński, K.; Radecki-Pawlik, A.; Gharabaghi, B. Scour depth model for grade-control structures. *J. Hydroinform.* **2018**, *20*, 117–133. [CrossRef]

29. Plesiński, K.; Radecki-Pawlik, A.; Wyżga, B. Sediment Transport Processes Related to the Operation of a Rapid Hydraulic Structure (Boulder Ramp) in a Mountain Stream Channel: A Polish Carpathian Example. In *Sediment Metters*; Heininger, P., Cullmann, J., Eds.; Springer: Koblenz, Germany, 2015; pp. 39–58.

30. Plesiński, K.; Pachla, F.; Radecki-Pawlik, A.; Tatara, T.; Radecki-Pawlik, B. Numerical 2D simulation of morphological phenomena of a block ramp in Poniczanka Stream: Polish Carpathians. In Proceedings of the 7th IAHR International Symposium on Hydraulic Structures, Aachen, Germany, 15–18 May 2018; Bung, D., Tullis, B., Eds.; Utah State University: Logan, UT, USA, 2018.

31. Radecki-Pawlik, A.; Plesiński, K.; Radecki-Pawlik, B.; Kuboń, P.; Manson, R. Hydrodynamic parameters in a flood impacted boulder block ramp: Krzczonówska mountain stream, Polish Carpathians. *J. Mt. Sci.* **2018**, *15*, 2335–2346. [CrossRef]

32. Kałuża, T.; Radecki-Pawlik, A.; Szoszkiewicz, K.; Plesiński, K.; Radecki-Pawlik, B.; Laks, I. Plant basket hydraulic structures (PBHS) as a new river restoration measure. *Sci. Total Environ.* **2018**, *627*, 245–255. [CrossRef] [PubMed]

33. Wyżga, B.; Oglęcki, P.; Radecki-Pawlik, A.; Skalski, T.; Zawiejska, J. Hydromorphological complexity as a driver of the diversity of benthic invertebrate communities in the Czarny Dunajec River, Polish Carpathians. *Hydrobiologia* **2012**, *696*, 29–46. [CrossRef]

34. Wyżga, B.; Oglęcki, P.; Hajdukiewicz, H.; Zawiejska, J.; Radecki-Pawlik, A.; Skalski, T.; Mikuś, P. Interpretation of the invertebrate-based BMWP–PL index in a gravel-bed river: Insight from the Polish Carpathians. *Hydrobiologia* **2013**, *712*, 71–88. [CrossRef]

35. Wyżga, B.; Amirowicz, A.; Oglęcki, P.; Hajdukiewicz, H.; Radecki-Pawlik, A.; Zawiejska, J.; Mikuś, P. Response of fish and benthic invertebrate communities to constrained channel conditions in a mountain river: Case study of the Biała, Polish Carpathians. *Limnologica* **2014**, *46*, 58–69. [CrossRef]

36. Agoramoorthy, G.; Minna, J.H. Small size, big potential: Check dams for sustainable development. *Environ. Sci. Policy Sustain. Dev.* **2008**, *50*, 22–35. [CrossRef]

37. Conesa-Garcia, C.; Lenzi, M.A. *Check Dams, Morphological Adjustments and Erosion Control in Torrential Streams*; Nova Science Publishers: New York, NY, USA, 2010; ISBN 978-1-61761-749-2.

38. Renganayaki, P.; Elango, L. A review on managed aquifer recharge by check dams: A case study near Chennai, India. *Int. J. Res. Eng. Technol.* **2013**, *2*, 416–423. [CrossRef]

39. Castillo, C.; Pérez, R.; Gómez, J.A. A conceptual model of check dam hydraulics for gully control: Efficiency, optimal spacing and relation with step-pools. *Hydrol. Earth Syst. Sci.* **2014**, *18*, 1705–1721. [CrossRef]

40. Bucala, A.; Radecki-Pawlik, A. Wpływ regulacji technicznej na zmiany morfologii górskiej potoku: Potok Jamne, Gorce, The influence of hydrotechnical structures on morphological changes of Jamne stream channels in the Gorce Mountains, Polish Carpathians. *Acta Sci. Pol. Form. Circumiectus* **2011**, *10*, 3–16.

41. Horton, R.E. Erosional development of streams and their drainage basins; hydrophysical approach to quantitative morphology. *Bull. Geol. Soc. Am.* **1945**, *56*, 275–370. [CrossRef]

42. Kajetanowicz, Z. Untersuchungsmethoden der Sinkstoffbewegung in den Flüssen Südpolens. In Proceedings of the VI Baltische Hydrologische Konferenz, Berlin, Germany, August 1938; Volume 19D, p. 19.

43. Chalov, S.; Golosov, V.; Tsyplenkov, A.; Theuring, P.; Zakerinejad, R.; Märker, M.; Samokhin, M.A. Toolbox for sediment budget research in small catchments. *Geogr. Environ. Sustain.* **2017**, *10*, 43–68. [CrossRef]

44. Radecki-Pawlik, A.; Bucała, A.; Plesiński, K.; Oglęcki, P. Ecohydrological conditions in two catchments in the Gorce Mountains: Jaszcze and Jamne streams—Western Polish Carpathians. *Ecohydrol. Hydrobiol.* **2014**, *14*, 229–242. [CrossRef]

45. Meyer-Peter, E.; Mueller, R. Formulas for bedload transport. In Proceedings of the II Congress IAHR 1948, Stockholm, Sweden, 7 June 1948; pp. 39–64.

46. *Wallingford University the SandCalc1_1 Software*; Wallingford University: Wallingford, UK, 1996.

47. Frost, J.; Clarke, R.T. Estimating the T-tear flood by the extension of records of partial duration series. *Hydrol. Sci. J.* **1972**, *17*, 209–217. [CrossRef]

48. Punzet, J. Empiryczny system ocen charakterystycznych przepływów rzek i potoków w karpackiej części dorzecza Wisły (in Polish). *Wiadomości Inst. Meteorol. I Gospod. Wodnej* **1981**, *7*, 31–40.

49. Radecki-Pawlik, A. Woda-v. 2. 0—A simple hydrological computer model to calculate the T-year flood. Hydrological processes in the catchment. Cracow University of Technology. In Proceedings of the International Conference 1995, Cracow, Poland, 26–28 October 1995.

50. Florkowski, J.; Adamek, D. Wielozadaniowy zbiornik retencyjny Młynne na rz. Łososinie (in Polish), the multi-role storage reservoir Mlynne on the river Lososina. *Gospod. Wodna* **2004**, *5*, 197–203.

51. HYDROPROJEKT Designing Office. *Studium do Założeń Kompleksowych Gospodarki Wodnej w Dorzeczu Łososiny*; CBS i PBW Hydroprojekt: Kraków, Poland, 1953. (In Polish)

52. HYDROPROJEKT Designing Office. *Założenia Gospodarki Wodnej Dorzecza Dunajca*; CUGW, CBS i PBW Hydroprojekt: Kraków, Poland, 1961. (In Polish)

53. Kaszowski, L.; Kotarba, A. *Wpływ Katastrofalnych Wezbrań na Przebieg Procesów Fluwialnych*; Prace Geograficzne Instytutu Geografii PAN: Warszawa, Poland, 1970. (In Polish)

54. Gorczyca, E. *Przekształcanie Stoków Fliszowych Przez Procesy Masowe Podczas Katastrofalnych Opadów (Dorzecze Łososiny)*; Wydawnictwo Uniwersytetu Jagiellońskiego: Cracow, Poland, 2005; ISBN 83-233-1957-X. (In Polish)

55. Shvidchenko, A.B.; Kopaliani, Z.D. Hydraulic modeling of bed load transport in gravel-bed Laba River. *J. Hydraul. Eng.* **1998**, *124*, 778–785. [CrossRef]

56. Lisle, T.E.; Nelson, J.M.; Pitlick, J.; Madej, M.A.; Barket, B.L. Variability of bed mobility in natural, gravel-bed channels and adjustments to sediment load at local and reach scales. *Water Resour. Res.* **2000**, *36*, 3743–3755. [CrossRef]

water

MDPI

Article

Distribution and Potential Risk of Heavy Metals in Sediments of the Three Gorges Reservoir: The Relationship to Environmental Variables

Lei Huang [1,2], Hongwei Fang [1,*], Ke Ni [1], Wenjun Yang [3,*], Weihua Zhao [3], Guojian He [1], Yong Han [1] and Xiaocui Li [1]

[1] State Key Laboratory of Hydro-science and Engineering, Department of Hydraulic Engineering, Tsinghua University, Beijing 100084, China; huangl05@mails.tsinghua.edu.cn (L.H.); nik13@mails.tsinghua.edu.cn (K.N.); heguojian@tsinghua.edu.cn (G.H.); hanyong_93@163.com (Y.H.); lixc17@mails.tsinghua.edu.cn (X.L.)
[2] State Key Laboratory of Lake Science and Environment, Nanjing 210008, China
[3] Changjiang River Scientific Research Institute of Changjiang Water Resources Commission, Wuhan 430015, China; zwh820305zwh@163.com
* Correspondence: fanghw@tsinghua.edu.cn (H.F.); yangwj@mail.crsri.cn (W.Y.); Tel.: +86-10-62781750 (H.F.); +86-27-82926353 (W.Y.)

Received: 11 November 2018; Accepted: 10 December 2018; Published: 12 December 2018

Abstract: In this study, surface sediment samples were taken from the Three Gorges Reservoir (TGR) in June 2015 to estimate the spatial distribution and potential risk of Cu, Zn, Cd, Pb, Cr, and Ni (34 sites from the mainstream and 9 sites from the major tributaries), and correlations with environmental variables were analyzed (e.g., median sediment size, water depth, turbidity, dissolved oxygen of the bottom water samples, and total organic carbon, total nitrogen, and total phosphorus of the surface sediment samples). Results show that the heavy metal concentrations in the sediments have increased over the last few decades, especially for Cd and Pb; and the sites in the downstream area, e.g., Badong (BD) and Wushan (WS), have had greater increments of heavy metal concentrations. The sampling sites from S6 to S12-WS are identified as hot spots for heavy metal distribution and have relatively high heavy metal concentrations, and there are also high values for the sites affected by urban cities (e.g., the concentrations of Zn, Cd, Cr and Ni for the site S12-WS). Overall, the heavy metal concentrations increased slightly along the mainstream due to pollutants discharged along the Yangtze River and sediment sorting in the reservoir, and the values in the mainstream were greater than those in the tributaries. Meanwhile, the heavy metal concentrations were generally positively correlated with water depth (especially for Ni), while negatively correlated with dissolved oxygen, turbidity, and median sediment size. These environmental variables have a great impact on the partition of heavy metals between the sediment and overlying water. According to the risk assessment, the heavy metals in the surface sediments of TGR give a low to moderate level of pollution.

Keywords: heavy metals; sediment; environmental variables; risk assessment; Three Gorges Reservoir

1. Introduction

Heavy metals exert significant negative impacts on the environment due to their abundance, persistence, and toxicity, which have been widely concerned by researchers [1–3]. Sediment particles, especially fine sediment particles, have a strong affinity to heavy metals in natural waters due to their specific surface area and surface active functional groups [4–6]. Thus, most heavy metal ions are adsorbed by sediments and transported in the particulate phase, with only a small portion remaining dissolved in the water column [7,8]. The accumulation of heavy metals at the bed surface, together with sediment, would result in a major source of heavy metals, which may be released into the overlying water under certain disturbances, posing a potential risk to the safety of the aquatic system [9,10]. Therefore, it is necessary to accurately assess the distribution and potential risk of heavy metals in the sediments.

The distribution of heavy metals in the sediments is affected by factors such as pollutant emissions, hydrodynamic conditions, sediment transport, and other physical and chemical processes [2,11,12]. Recently, human activities have exerted significant impacts on river systems. Firstly, pollutant effluents have greatly increased with the development of social economy, resulting in more heavy metals released into the aqueous systems [13]. Meanwhile, it is worth noting that many reservoirs have been built in rivers worldwide during the last decades [14], which operate to support a variety of social, economic, and ecological purposes. However, a reservoir operation would also alter the hydrological regime [15], accelerate sediment deposition and sorting along the main channel [16,17], and accordingly affect the occurrence and distribution of sediment-associated heavy metals [18], i.e., influencing the sediment and heavy metal balances in the river system [19–21]. Thus, the relationship between heavy metal distribution and environmental variables should be further studied due to human activities such as reservoir operation.

In this study, the distribution and potential risk of heavy metals (Cu, Zn, Cd, Pb, Cr, and Ni, which are the major heavy metals of concern) in the sediments of Three Gorges Reservoir (TGR, the largest hydraulic project in the world) is studied as an example, and the relationship to environmental variables is discussed. There have been some studies focusing on heavy metal distribution in sediments of the mainstream or tributaries of the TGR [22–27], and Zhao et al. [28] reviewed the available literature published on the heavy metal concentrations of the TGR sediments. However, only a few sampling sites were adopted by these studies, and the sampling sites were mostly localized or predominantly distributed in certain tributaries (e.g., the Daning River, Meixi River and Xiangxi River), which cannot characterize the heavy metal distribution in sediments of the whole reservoir well. Meanwhile, the relationship between heavy metal distribution and environmental variables have been generally lacking. Thus, a comprehensive sampling and analysis are conducted in this study, including 34 sites from the mainstream and 9 sites from the major tributaries, which is expected to provide a reference for the management of TGR and other similar reservoirs.

2. Materials and Methods

2.1. Study Area

Three Gorges Reservoir is located in the upstream Yangtze River, as shown in Figure 1. It started impoundment in June of 2003, and the fore-bay water level first reached its normal pool level (NPL) of 175 m in October, 2010. The total capacity of the reservoir is 393×10^8 m^3, with a flood control capacity of 221.5×10^8 m^3; and the surface area is 1084 km^2 under the NPL [29]. The reservoir region belongs to the Chongqing city and Hubei province, including the counties of Changshou (CS), Fuling (FL), Fengdu (FD), Zhongxian (ZX), Wanxian (WX), Yunyang (YY), Fengjie (FJ), Wushan (WS), and Badong (BD). According to the environmental and ecological monitoring bulletins of the TGR area [13], the population in the TGR area was 14.65 million by the end of 2015, including 13.17 million in Chongqing and 1.48 million in Hubei. In 2015, the gross domestic product (GDP) of the whole area was close to 700 billion CNY (China Yuan), i.e., an increase of 11.1% compared with that in 2014. Correspondingly, there were 212 million tons of industrial wastewater, and 815 million tons of urban domestic sewage discharged in the TGR area. Moreover, about 410,000 hectares of land are used for agriculture, with pesticide use of 601.8 tons and chemical fertilizer use of 135,000 tons. The land use in the region of TGR is presented in Figure S1. There is still a certain amount of sewage discharged by ships. Meanwhile, increasing the water level due to the operation of TGR results in decreasing flow velocity and increasing sediment deposition in the reservoir. The average flow velocity decreased from

1.33 m/s in Chongqing to 0.22 m/s in Badong in 2015 [13]. The sediment delivery ratio of TGR was estimated to be 13.3% in 2015, with most sediment deposited during the period from June to September and in the wide valley segments; and the sediment delivery ratio from June 2003 to December 2015 was 24.2% [30].

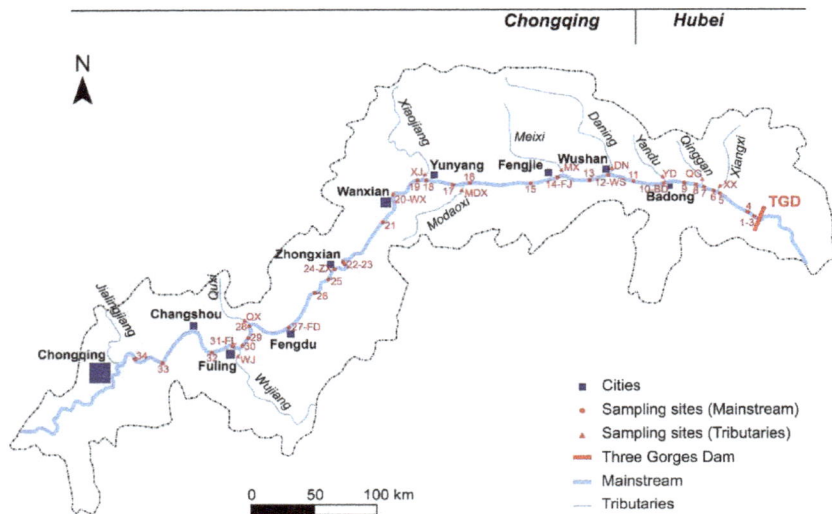

Figure 1. The study area and sampling sites in the Three Gorges Reservoir, including the sites in the mainstream and from the tributary estuaries (the size of square reflects the population of the city). BD—Badong, WS—Wushan, FJ—Fengjie, WX—Wanxian, ZX—Zhongxian, FD—Fengdu, FL—Fuling; and XX—Xiangxi, QG—Qinggan, YD—Yandu, DN—Daning, MX—Meixi, MDX—Modaoxi, XJ—Xiaojiang, QX—Quxi, WJ—Wujiang.

2.2. Sampling

During the period of 5–13 June 2015, 43 surface sediment samples were collected using a grab sampler from the TGR, when the fore-bay water level varied from 150 to 151 m. Figure 1 shows the distribution of these sampling sites, and the detailed latitude and longitudes are listed in Table S1 (see Supplementary Materials). There are 34 sites distributed in the mainstream from the dam to Chongqing with an average interval of 15–20 km, which covers the whole reservoir of about 600 km. Here the sites affected by urban cities are specially annotated, e.g., S10-BD represents the site affected by Badong. There are 9 more sites distributed in the major tributaries, including Xiangxi (XX), Qinggan (QG), Yandu (YD), Daning (DN), Meixi (MX), Modaoxi (MDX), Xiaojiang (XJ), Quxi (QX), and Wujiang (WJ). As the heavy metal concentrations in the sediments of tributaries generally show an increasing trend along the flow direction [28], the sampling sites of these tributaries are arranged in the tributary estuaries, see Figure 1. The collected sediment samples were stored in clean polyethylene bags and treated immediately once returning to the laboratory. The sediment was air dried, ground, and the impurities were removed through a 100-size sieve. Meanwhile, the bottom water samples were also collected just above the bed surface using a column sampler for the measurement of environmental variables.

2.3. Analytical Methods

The total heavy metal concentrations of Cu, Zn, Cd, Pb, Cr, and Ni in the sediments were measured using inductively coupled plasma-mass spectrometry (ICP-MS) as suggested by Liu et al. [31], i.e., the sediment samples were digested using distilled HF + HNO_3 solutions in screw-top Teflon beakers, and then used for the determination of heavy metal concentrations by ICP-MS. For more details, refer to Gao et al. [24]. Precision and accuracy were verified using standard reference material GBW07310 (GSD-10) purchased from the National Center of Reference Material (NCRM), and the average recoveries were 93–108%. Meanwhile, the chemical properties of sediments, including the total organic carbon (TOC), total nitrogen (TN), and total phosphorus (TP), were determined according to the standard methods for soil analysis [32], and the total polycyclic aromatic hydrocarbons (PAHs) and phthalic acid esters (PAEs) were determined using gas chromatography-mass spectrometry (GC-MS) analysis [33]. The grain size was measured by a laser scattering particle size distribution analyzer (LA-920, Horiba, Kyoto, Japan).

Moreover, the turbidity and dissolved oxygen (DO) of the bottom water samples were assessed in the field by the YSI meter (YSI Inc., Yellow Springs, OH, USA), and the corresponding concentrations of Cu, Zn, Cd, Pb, and Cr were assessed in the laboratory following the standard analytical methods [34]. The water depth, H, was detected with an ultrasonic wave detector.

2.4. Risk Assessment

The potential ecological risk was used to assess the heavy metal eco-risk in the sediments [35]:

$$E_r^i = T_r^i \cdot C_f^i, \tag{1}$$

where T_r^i is the toxic response factor of each heavy metal, i.e., Cu = 5, Zn = 1, Cd = 30, Pb = 5, Cr = 2, and Ni = 5; C_f^i is calculated as $C_f^i = C_s^i/C_n^i$, where C_s^i is the measured metal concentration in the sediments, and C_n^i is the regional background value. Here the background value in the Yangtze River was used, i.e., 35, 78, 0.25, 27, 82, and 33 mg/kg for Cu, Zn, Cd, Pb, Cr, and Ni, respectively [28,36]. The ecological risk of each heavy metal was classified as low ($E_r^i < 40$), moderate ($40 \leq E_r^i < 80$), considerable ($80 \leq E_r^i < 160$), high ($160 \leq E_r^i < 320$), or very high ($E_r^i \geq 320$). The comprehensive index, R_I, of potential ecological risk is expressed as

$$R_I = \sum_{i=1}^{n} E_r^i, \tag{2}$$

and the ecological risk level of all heavy metals is defined as low ($R_I < 150$), moderate ($150 \leq R_I < 300$), considerable ($300 \leq R_I < 600$), or very high ($R_I > 600$).

2.5. Statistical Analysis

The principal component analysis (PCA), hierarchical cluster analysis (HACA), and Pearson correlation analysis (CA) were conducted using SPSS 20.0. The Kaiser-Meyer-Olkin (KMO) and Bartlett's test were introduced to evaluate the validity of PCA [2]. Moreover, the redundancy analysis (RDA) was executed using Canoco 4.5 to analyze the interactions between the heavy metal distribution and the relevant environmental variables.

3. Results

3.1. Environmental Variables

The physical and chemical parameters of the sampling sites are listed in Table S1, including the turbidity, DO of the bottom water samples, the water depth, median sediment size (D_{50}), textural composition (i.e., clay: $D \leq 0.004$ mm; silt: 0.004–0.062 mm; and sand: 0.062–2.0 mm), and the TOC, TN, TP, and total PAHs and PAEs of the surface sediment samples. Figure 2a presents the textural composition of the surface sediment samples, corresponding to a clay content of 25–50%, a silt content of 50–75%, and a sand content of 0–25%, which can be defined as clayey silt following Shepard [37]. Particularly, the sampling site S10-BD has a greater clay content of 57.01%, while S33 has a relatively greater sand content of 24.71%. Figure 2b shows the variation of median sediment size with distance to the dam. The overall range of D_{50} is from 0.004 to 0.020 mm, and there is a decreasing trend of D_{50} when approaching the dam (R^2 = 0.68), indicating a significant sediment sorting along the main channel.

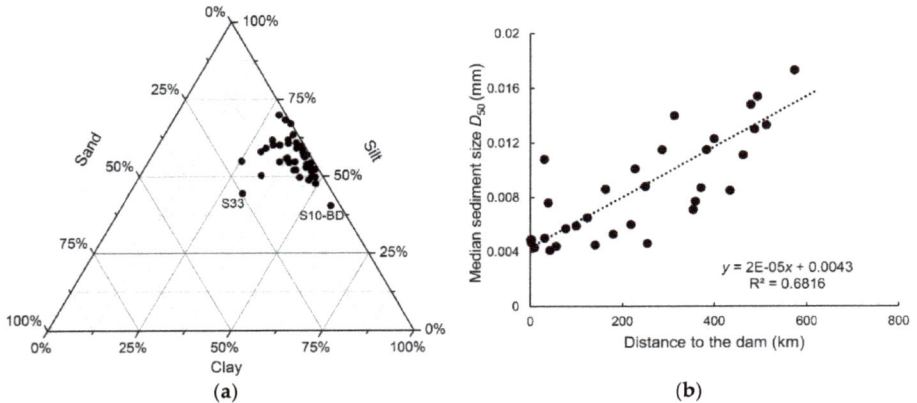

Figure 2. (**a**) Textural composition of the surface sediment samples in the Three Gorges Reservoir (triangle diagram), and (**b**) the variation of median sediment size D_{50} with the distance to the dam.

The box plots of other environmental variables are further presented in Figure 3. The average turbidity of these bottom water samples was 23.14 NTU. The measured DO concentration was 7.23 ± 0.62 mg/L (i.e., an oxidized status), implying that the TGR is an un-stratified reservoir with essentially uniform oxygen concentration from the water surface to the bed sediments, thus maintaining oxic conditions at the sediment surface [19,38]. The water depth ranges from 5 to 115 m, which can represent the different water depths in the reservoir well. The TOC, TN and TP indicate the trophic status of the sediment, and they were 1.64 ± 0.36%, 951.7 ± 34.1 mg/kg and 963.2 ± 273.0 mg/kg, respectively. The organic pollutants of the total PAHs and PAEs were estimated to be 935.9 ± 311.4 ng/g and 1740.6 ± 1181.5 ng/g, respectively. Apparently, the total PAEs exhibited a greater variability.

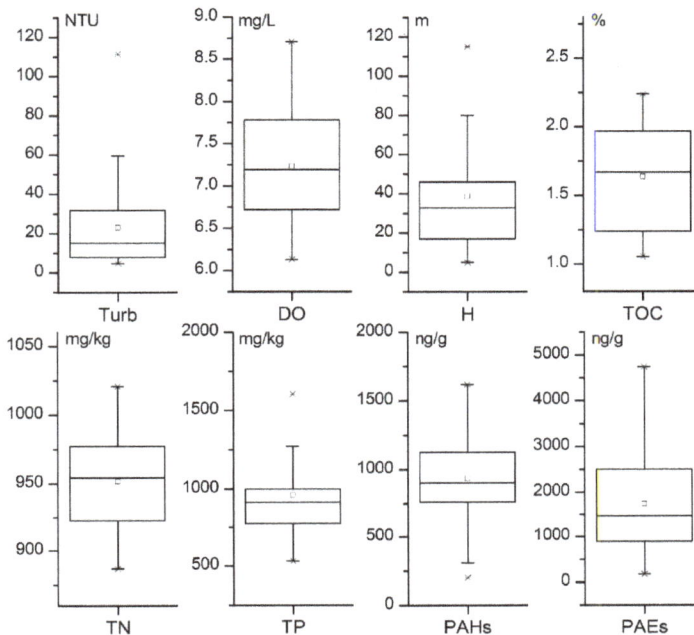

Figure 3. Box plots of the measured physical and chemical environmental variables, including the turbidity, dissolved oxygen (DO), and water depth (*H*) of the bottom water samples, and the total organic carbon (TOC), total nitrogen (TN), total phosphorus (TP), and the total polycyclic aromatic hydrocarbons (PAHs) and phthalic acid esters (PAEs) of the surface sediment samples.

3.2. Heavy Metal Concentrations

The measured heavy metal concentrations in the surface sediment samples are listed in Table 1, where the results of the mainstream and tributaries are separately presented. Meanwhile, the measured values in previous studies of the TGR are also listed in Table 1. Overall, the heavy metal concentrations in the sediments of mainstream were relatively greater than those in the sediments of tributaries, which is consistent with Zhao et al. [28]. Tang [39] observed that the heavy metal concentration in the north bank of TGR was lower than that in the south bank due to the effects of tributaries, which also verifies that there are lower heavy metal concentrations in the tributaries than the mainstream. Meanwhile, the remarkable variation of Cd, i.e., 31.4% in the mainstream and 25.2% in the tributaries, reflects the influence of anthropogenic activity.

Compared with the previous results of the TGR before impoundment, i.e., TGR-1985 measured by Xu et al. [40], the average heavy metal concentrations has mostly increased over the last few decades. A more detailed comparison for the sites affected by urban cities are further presented in Table S2. There was a significant increase in heavy metal concentrations of these sites, especially for the heavy metals Cd and Pb; and the sites in the downstream area had greater increments of heavy metal concentrations, e.g., Badong and Wushan. Meanwhile, it is worth noting that the surface sediment samples in Xu et al. [40] were collected from the riverside, which is more likely to be affected by the urban cities. After impoundment, there was a tendency for the heavy metal concentrations to still slightly increase. Moreover, if compared with the soil standards (GB15618-1995) [41], most of these sites can be classified into category II, i.e., a low contamination.

Table 1. Comparison of heavy metal concentrations with previous studies (unit: mg/kg).

Sampling Sites		Cu	Zn	Cd	Pb	Cr	Ni	References
Mainstream	TGR-1985 (N = 17)	62.5	160.6	0.27	25.7	145.1	36.9	[40]
	TGR < 2005 (N = 126)	53.52	146.8	0.605	50.84	87.15	37.11	[39]
	TGR-2014 (N = 24)	54.2	174	0.878	51	86.8	42.7	[22,42]
	A review	60.82 ± 28.07	148.11 ± 60.84	0.63 ± 0.81	42.73 ± 21.73	125.56 ± 71.97	42.31 ± 7.43	[28]
Tributaries	TGR-2008 (N = 24)	76.03	137.63	0.75	59.4	86.31	46.81	[43]
	TGR-2010 (N = 73)	56.4	130.3	0.9	44	84.9	45.7	[24]
	A review	53.31 ± 25.88	129.16 ± 75.92	0.72 ± 0.67	42.93 ± 23.06	79.28 ± 28.60	42.45 ± 8.80	[28]
Mainstream (N = 34)	Mean ± SD	61.00 ± 15.04 (24.7%) *	151.63 ± 26.40 (17.4%)	0.92 ± 0.29 (31.4%)	55.38 ± 10.45 (18.9%)	101.43 ± 18.99 (18.7%)	43.00 ± 8.63 (20.1%)	This study
	Range	35.26–96.34	106.73–204.83	0.61–2.10	40.90–83.06	70.32–171.12	27.88–55.89	
Tributaries (N = 9)	Mean ± SD	52.92 ± 14.45 (27.3%)	138.34 ± 20.37 (14.7%)	0.86 ± 0.22 (25.2%)	48.19 ± 9.85 (20.4%)	92.98 ± 7.76 (8.4%)	44.86 ± 3.03 (6.8%)	
	Range	34.25–86.37	99.08–180.06	0.48–1.30	28.77–67.81	84.39–111.46	39.54–50.65	
Soil standards (GB 15618–1995)	I	35	100	0.2	35	90	40	[41]
	II (pH > 7.5)	100	300	0.6	350	350	60	
	III	400	500	1.0	500	400	200	

* the coefficient of variation (CV).

3.3. Spatial Distribution of Heavy Metals

Figure 4 shows the spatial distribution of Cu, Zn, Cd, Pb, Cr, and Ni in the TGR, and the sampling sites in the mainstream and tributaries are separately presented in each figure. The red lines represent the polynomial trend lines, and the white lines indicate the soil standard values of GB 15618-1995 (i.e., category I, II, or III). Overall, the heavy metal concentrations increased slightly along the mainstream, especially the concentration of Ni. Firstly, there are pollutants of point and non-point sources discharged gradually along the Yangtze River (see Figure S1), resulting in a higher heavy metal concentration in the downstream compared with that in the upstream. Secondly, the operation of TGR leads to sediment sorting along the main channel (Figure 2b), i.e., fine sediment deposits close to the dam, while coarse sediment deposits at the reservoir tail. So the strong affinity of fine sediment to the pollutants also results in a higher heavy metal concentration downstream.

Meanwhile, the sampling sites from S6 to S12-WS were found to have relatively high heavy metal concentrations, i.e., hot spots for the distribution of heavy metal concentration. Whereas, there was relatively lower heavy metal concentrations for the sampling sites from S25 to S30, indicating that local pollutant emissions have a great influence on the spatial distribution of heavy metal concentrations. Moreover, the sites affected by urban cities also have relatively high heavy metal concentrations. For example, the concentrations of Cd and Cr for the site S10-BD (Badong), the concentrations of Zn, Cd, Cr and Ni for the site S12-WS (Wushan), the concentration of Cr for the site S24-ZX (Zhongxian), and the concentrations of Zn, Cd, and Cr for the sites S34 (probably affected by the main urban districts of Chongqing) are much greater than other sites. However, the urban emission of pollutants can only affect a certain range, and the pollutants will then deposit onto the riverbed together with the sediment [44]. Similarly, the tributaries exhibited lower heavy metal concentrations than the mainstream, except for the site from Quxi, which had relatively high concentrations of all six heavy metals.

As compared with the soil standard values, the concentrations of most heavy metals can be classified into category II (i.e., a low contamination), while the Cd concentrations in eight sites belong to category III, that is close to the high contamination threshold value, including S4, S9, S10-BD, S12-WS, S32, S33, S34, and QX. The spatial distribution of Cd seems to be slightly different from other heavy metals, which will be further discussed in the following sections.

4. Discussion

4.1. Source Identification

The PCA approach was performed to identify the characteristics (or sources) of the heavy metals in the sediments of TGR [45], and $a > 0.5$ of KMO (0.595) and significant Bartlett's test (<0.001) demonstrated its validity. Figure 5 shows the relationships among these heavy metals. Two principal components are extracted with eigenvalues greater than 1, which can explain 82.8% of the total variance.

As shown in Figure 5, component 1 that explains 49.2% of the variance mainly represents the pollution of Ni, Cu, and Pb, while component 2 that explains 33.6% of the variance mainly represents the pollution of Cd. These results imply a similar source for the heavy metals Ni, Cu, and Pb, which is different from that of Cd. Based on the Pb isotope composition, Bing et al. [22] concluded that Pb in the sediments of TGR mainly originated from industrial discharge, domestic sewage, mining, smelting, and the shipping industry. However, Cd is an identification element of agricultural activities, i.e., the application of pesticides and fertilizers [25,46]. Liu et al. [47] stated that Cd is found predominantly in phosphate fertilizers as an impurity of phosphate rocks; and the amount of pesticides and fertilizers used in the TGR were 601.8 tons of 135,000 tons, respectively, in 2015 [13]. Meanwhile, there is also a high content of Cd in coal mines [48], and Liu et al. [49] concluded that Cd-rich coal mining activity may contribute to the high concentration of Cd in the Three Gorges region.

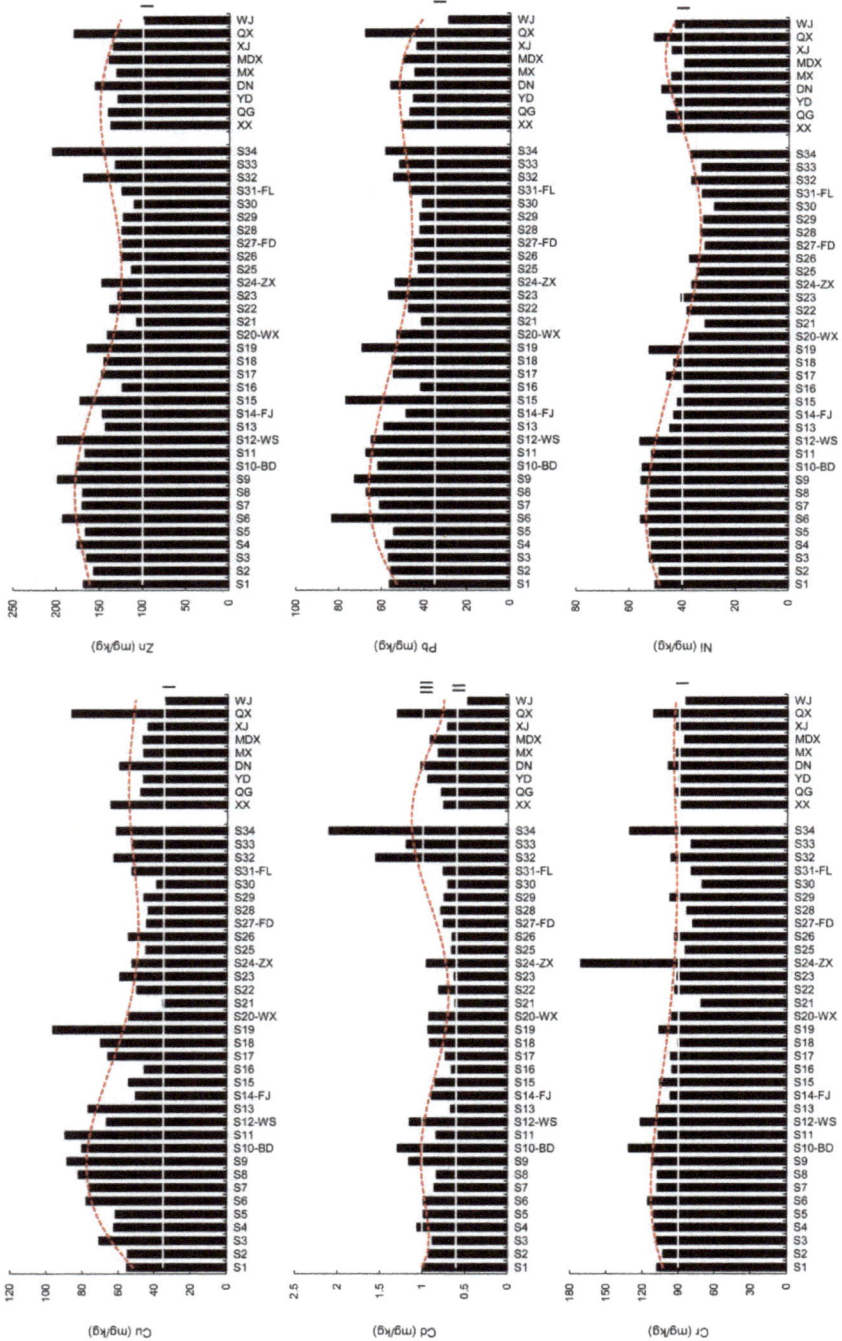

Figure 4. Spatial distribution of Cu, Zn, Cd, Pb, Cr, and Ni in the Three Gorges Reservoir. The white lines represent the standard values of the soil (GB 15618-1995).

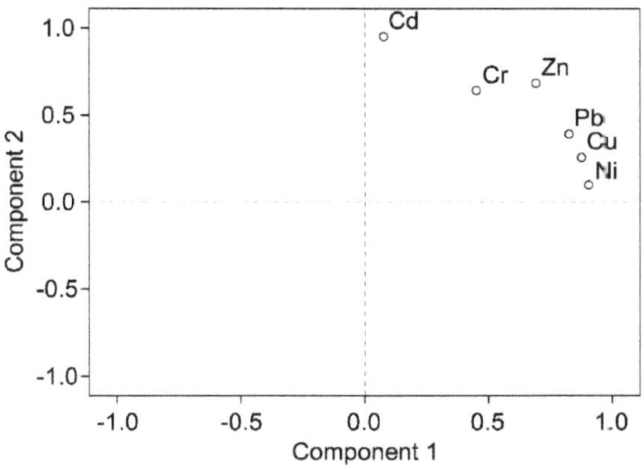

Figure 5. Plot of loading of two principal components in the rotated space.

Moreover, the CA analysis was conducted based on the PCA results, as shown in Figure 6a. Accordingly, the sampling sites can be classified into three main groups: (1) PC1 < 0 and PC2 < 1 with a low pollution of these heavy metals, including S14-FJ, S16, S20-WX, S21, S22, S25, S26, S27-FD, S28, S29, S30, S31-FL, S33, and QG, YD, MX, MDX, XJ, WJ; (2) 0 < PC1 < 2 with different degrees of Ni, Cu, and Pb pollution, including S1, S2, S3, S4, S5, **S6**, S7, **S8**, **S9**, S10-BD, **S11**, S12-WS, S13, S15, S17, S18, **S19**, S23, and XX, DN, QX; and (3) PC2 > 1 mainly affected by Cd, including S24-ZX, S32, and S34, where the site S24-ZX also has a high content of Cr. For the mainstream, the sites are mainly classified following the locations, i.e., the upstream sites exhibit a lower pollution compared with the downstream sites, which is also shown in Figure 4. Among the downstream sites, S6, S8, S9, S11, and S19 exhibit relatively high pollution. Moreover, the sites from most tributaries are light polluted, except for the sites XX, DN, and QX which exhibit different degrees of Ni, Cu, and Pb pollution.

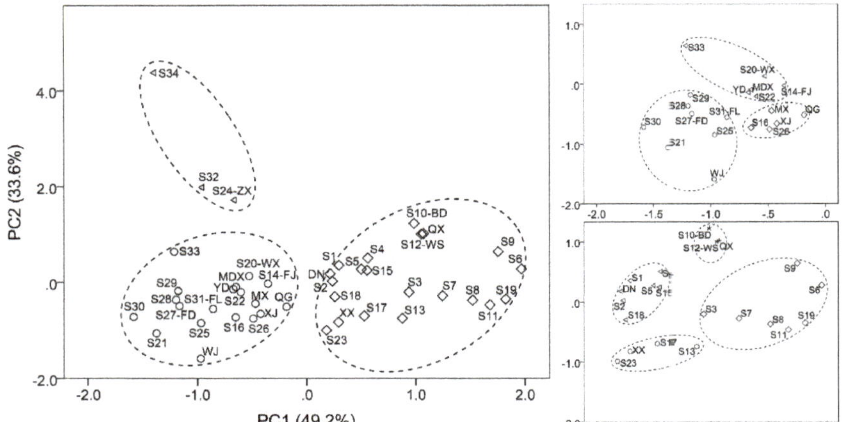

(a)

Figure 6. *Cont.*

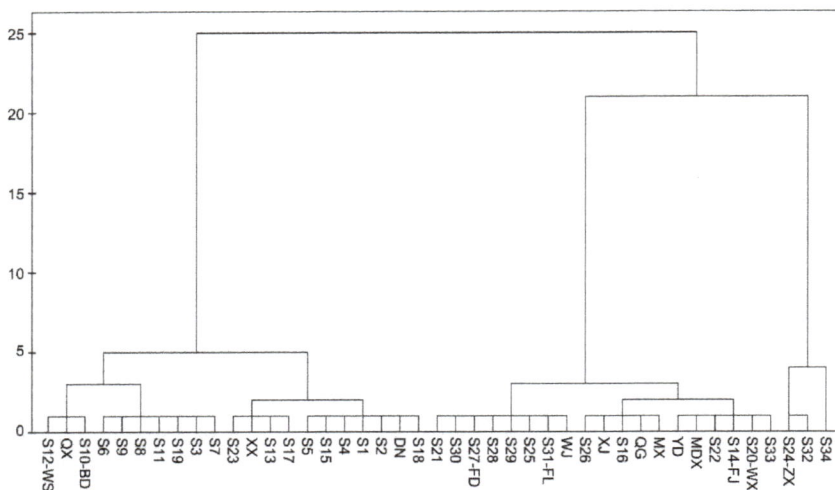

(b)

Figure 6. (a) Principal component analysis (PCA) results and (b) hierarchical diagram of the sampling sites.

In Figure 6a, the groups 1 and 2 are further divided into several sub-groups according to the Hierarchical diagram (see Figure 6b). For example, the sites S21, S30, and WJ in group 1 are the least polluted, and S33 also exhibits a certain degree of Cd pollution; the sites S6, S9, and S19 in group 2 have relatively high concentrations of Ni, Cu, and Pb compared with other sites, and S10-BD, S12-WS, and QX have medium to high concentrations of all these heavy metals.

4.2. Impacts of Environmental Variables

Table 2 shows a correlation matrix between the six heavy metals and environmental variables described in Section 3.1. Overall, the heavy metal concentrations are negatively correlated with the D_{50}, turbidity, and DO, while positively correlated with the water depth, H, and TOC. The redox condition can affect the solubility of heavy metals, which will be more likely fixed in the sediment under a reduction condition, i.e., a low DO condition [50]. Fine sediment particles have a higher affinity for the heavy metals. Xiao et al. [51] found that the sediment in Xiangxi is mainly comprised of chlorite, illite, and quartz, and the heavy metal concentration increases with decreasing sediment size and increasing content of chlorite and illite. Thus, in the region with a great water depth (e.g., the region close to the dam), a relatively low DO value and fine sediment size due to sediment sorting will lead to a greater heavy metal concentration in the sediment. Meanwhile, as organic matter can stabilize heavy metals in the sediment [50], the heavy metal concentrations are also positively correlated with the TOC values.

Table 2. Correlation analysis between the heavy metal concentration and environmental variables.

	D_{50}	Turb	DO	H	TOC	TN	TP	PAHs	PAEs
Cu	−0.319 *	−0.268	−0.452 **	0.253	0.172	−0.086	−0.156	0.298	0.314
Zn	−0.222	−0.182	−0.450 **	0.351 *	0.453	−0.074	−0.120	0.305	0.362 *
Cd	0.318 *	0.007	−0.141	−0.010	0.134	0.002	0.176	0.463 **	0.437*
Pb	−0.336 *	−0.042	−0.464 **	0.333 *	0.359	−0.086	−0.116	0.321	0.351 *
Cr	−0.196	−0.250	−0.190	0.171	−0.212	0.158	0.141	0.251	0.351 *
Ni	−0.664 **	−0.542 **	−0.559 **	0.564 **	0.491	−0.178	−0.443	0.108	0.221

* significant correlation at the 0.05 level (2-tailed); ** significant correlation at the 0.01 level (2-tailed).

Particularly, there is a significant negative correlation between the heavy metal Ni and the D_{50}, turbidity and DO, and a significant positive correlation between Ni and H ($p < 0.01$), i.e., the Ni concentration increases along the mainstream, implying that Ni mainly originates from upstream. Meanwhile, there are also significant negative correlations between DO and the heavy metals Cu, Zn and Pb ($p < 0.01$), and between D_{50} and Cu and Pb ($p < 0.05$), i.e., there is a similar source of Cu, Zn and Pb to that of Ni. However, a significant positive correlation is observed between D_{50} and Cd ($p < 0.05$), implying a relatively different source of Cd compared to other heavy metals, which is also shown in Figure 5. As previously stated, Cd generally serves as an impurity of phosphate rocks, so it is slightly positively correlated with the TP.

Moreover, the TN and TP are not significantly correlated with the heavy metals, implying different sources. However, total PAH is positively correlated with Cd ($p < 0.01$), and total PAE is also significantly correlated with Zn, Cd, Pb and Cr ($p < 0.05$), implying similar sources and transport characteristics.

The RDA analysis was conducted to further investigate the influences of environmental variables on heavy metal distributions, as shown in Figure 7. The lengths of the environmental variable arrows reflect the degree of relevance, and it can be found that there are greater correlations between the heavy metal concentrations and H, DO, turbidity, and D_{50}, which can also be concluded from Table 2. Meanwhile, according to angles between these arrows (i.e., the projections), the heavy metal concentrations are positively correlated to the water depth, H, especially for Ni; while they are generally negatively correlated with DO, turbidity, and D_{50}. In comparison, the TOC, TN, TP, and total PAEs and PAHs have slight influences on the heavy metal concentration.

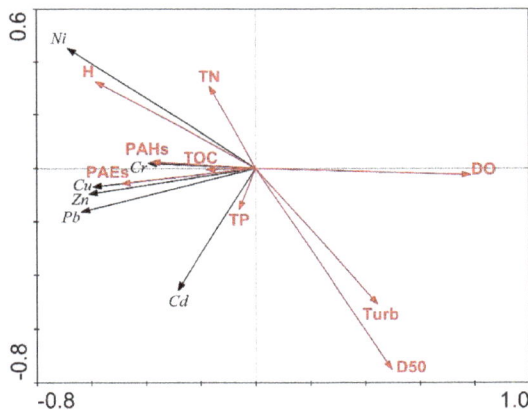

Figure 7. Redundancy analysis (RDA) of heavy metals and environment variables.

4.3. Risk Assessment

The risks of heavy metals in the sediments were assessed using the potential ecological risk index. According to the R_I results, the sampling sites S4, S6, S9, S10-BD, S12-WS, S32, S33, S34, and DN and QX exhibit a moderate potential ecological risk (i.e., $150 \leq R_I < 300$), and other sampling sites exhibit a low potential ecological risk, as shown in Figure 8a. The sampling site S34 had the largest value of R_I. The mainstream poses a greater potential ecological risk of heavy metals than the tributary [24], i.e., 140.49 and 130.40 for the average value of R_I, respectively. These results are in accordance with the metal distribution pattern in the sediment [2]. Moreover, the average E_r^i value of each heavy metal follows: Cd (109.02) > Pb (9.98) > Cu (8.47) > Ni (6.57) > Cr (2.43) > Zn (1.91), as shown in Figure 8b. Thus, Cd exhibits a considerable ecological risk, and other heavy metals exhibit low ecological risks. Overall, the heavy metals in the surface sediments of the TGR represent low to moderate pollution.

The bioavailability of heavy metals (i.e., the different fractions) should be further analyzed for the risk analysis, to determine the direct threat to the surrounding environment [22,52].

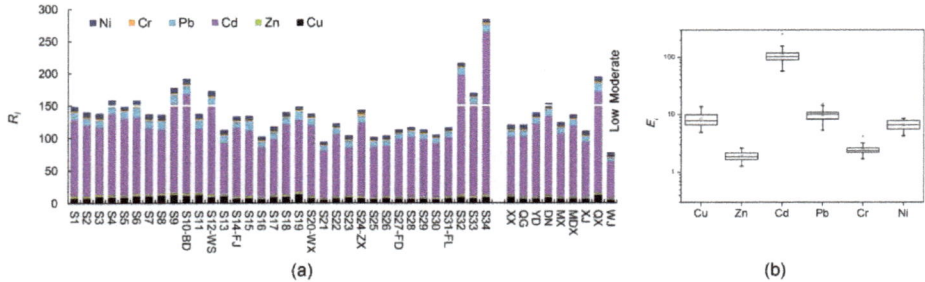

(a)　　　　　　　　　　　　　　　　　(b)

Figure 8. Risk assessment of heavy metals in the sediments. (**a**) the comprehensive index R_I of potential ecological risk (the white line represents the split line between the low and moderate ecological risk levels); and (**b**) the statistics of E_r^i for each heavy metal.

4.4. Partition of Heavy Metals

The heavy metal concentrations in the bottom water samples were measured for these sampling sites, and further analysis was conducted to compare the heavy metal concentration in the sediment and that in the overlying water. We define a variable of $K_d = C_s/C_w$ that reflects the partition of heavy metal between the sediment and overlying water, where C_s and C_w are the total heavy metal concentrations in the surface sediment and bottom water samples, respectively. The statistics of K_d for Cu, Zn, Cd, Pb, and Cr are shown in Figure 9, with the results of the mainstream and tributary presented separately.

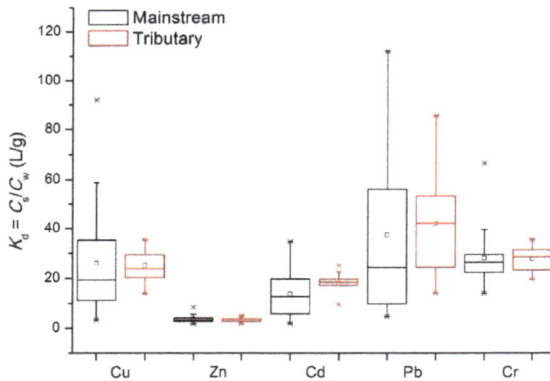

Figure 9. Statistical analysis of the heavy metal partition between the surface sediment and bottom water samples, with the sites from the mainstream and tributaries separately presented.

In Figure 9, different ranges of K_d values were obtained for different heavy metals. For comparison, there is a relatively small K_d for the heavy metal Zn, i.e., 3.59 ± 1.31; but the values for Pb, Cr, and Cu are relatively larger, i.e., 38.37 ± 32.31, 28.00 ± 9.18, 25.80 ± 19.11, respectively. Moreover, the median value of K_d for each heavy metal generally satisfies that: Tributary > Mainstream, i.e., the K_d values of the sampling sites in the tributaries bias toward a greater value (a distribution of right deviation). According to the definition, a greater K_d implies that there might be more pollutants existing in the sediment, while a smaller K_d represents relatively more pollutants in the overlying water.

Apparently, the K_d values exhibit significant variations for all heavy metals (especially in the mainstream with a coefficient of variation of 36–93%), indicating that it might be affected by the environmental variables, such as the D_{50}, turbidity, DO, H, and TOC as described above. Similarly, a correlation analysis was conducted to investigate the influences of these environmental variables on the K_d values, as shown in Table 3 and Figure 10. It was found that there is a more significant correlation between the K_d values and environmental variables than that between the heavy metal concentration and environmental variables, as compared with Table 2, indicating that these environmental variables have a greater impact on the partition of heavy metals between the sediment and overlying water. Overall, the K_d values for these heavy metals are significantly negatively correlated with the D_{50}, turbidity, and DO. The turbidity reflects the suspended sediment concentration in the overlying water, and a greater turbidity implies that there will be more heavy metals distributed in the overlying water, thus resulting in a smaller K_d value.

Table 3. Correlation analysis between the K_d values and environmental variables.

	D_{50}	Turb	DO	H	TOC
K_d_Cu	−0.511 **	−0.425 **	−0.397 **	0.295	0.124
K_d_Zn	−0.417 **	−0.375 *	−0.483 **	0.126	0.284
K_d_Cd	−0.642 **	−0.497 **	−0.539 **	0.520 **	0.372
K_d_Pb	−0.613 **	−0.422 **	−0.509 **	0.493 **	0.433
K_d_Cr	−0.332 *	−0.325 *	−0.154	0.103	−0.366

* significant correlation at the 0.05 level (2-tailed); ** significant correlation at the 0.01 level (2-tailed).

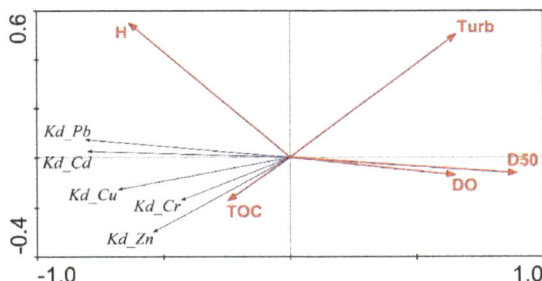

Figure 10. Redundancy analysis (RDA) analysis results of the K_d values and environment variables.

5. Conclusions

Heavy metals exert significant negative impacts on the aqueous environment due to their abundance, persistence and toxicity. The reservoir operation affects the distribution of heavy metals through changing the hydraulic regime of natural rivers, and causing sediment sorting and deposition in the reservoir, which leads to variations in water depth, median sediment size, dissolved oxygen and other environmental variables. In this study, the spatial distribution and potential risk of Cu, Zn, Cd, Pb, Cr, and Ni in the sediments of TGR were investigated as an example, which is also expected to provide references for the management of other similar reservoirs. The main conclusions are drawn as follows:

1. Heavy metal concentrations increase slightly along the mainstream due to pollutant emission and sediment sorting, and the sites from S6 to S12-WS are identified as hot spots for heavy metal distribution. Meanwhile, the heavy metal concentrations in the mainstream are relatively greater than those in the tributaries.

2. There is a similar source for the heavy metals Ni, Cu, and Pb, which is different from that of Cd. Meanwhile, the heavy metal concentrations are generally positively correlated to the water depth,

H, while negatively correlated with DO, turbidity, and D_{50}; and the environmental variables exert a greater impact on the heavy metal partition between the sediment and overlying water.

3. According to the risk assessment, the heavy metals in the surface sediments of TGR show a low to moderate pollution. The average E_i value of each heavy metal follows: Cd > Pb > Cu > Ni > Cr > Zn, where Cd exhibits a considerable ecological risk, and other heavy metals exhibit low ecological risks.

Supplementary Materials: The following are available online at http://www.mdpi.com/2073-4441/10/12/1840/s1, Figure S1: Land use in the region of Three Gorges Reservoir, Table S1: Location of the sampling sites and the measured environmental variables, Table S2: A more detailed comparison with the results of Xu et al. (1999) for the sites affected by urban cities.

Author Contributions: Conceptualization, H.F. and W.Y.; methodology, L.H. and W.Z.; formal analysis, G.H. and X.L.; investigation, L.H., K.N., W.Z. and Y.H.; writing—original draft preparation, L.H.; writing—review and editing, H.F.

Funding: This research was funded by the National Natural Science Foundation of China (91647210, 11802158, 51479213), 111 Project (No. B18031), National Key Research and Development Program of China (2016YFC0402506), Research Foundations of State Key Laboratory of Hydro-science and Engineering (2018-KY-03) and State Key Laboratory of Lake Science and Environment (2016SKL012), and Public Science and Technology Research Funds Projects of Ministry of Water Resources (No. 201501042).

Conflicts of Interest: The authors declare no conflicts of interest.

References

1. Benson, N.U.; Udosen, E.D.; Essien, J.P.; Anake, W.U.; Adedapo, A.E.; Akintokun, O.A.; Fred-Ahmadu, O.H.; Olajire, A.A. Geochemical fractionation and ecological risks assessment of benthic sediment-bound heavy metals from coastal ecosystems off the Equatorial Atlantic Ocean. *Int. J. Sediment Res.* **2017**, *32*, 410–420. [CrossRef]

2. Fu, J.; Zhao, C.P.; Luo, Y.P.; Liu, C.S.; Kyzas, G.Z.; Luo, Y.; Zhao, D.Y.; An, S.Q.; Zhu, H.L. Heavy metals in surface sediments of the Jialu River, China: Their relations to environmental factors. *J. Hazard. Mater.* **2014**, *270*, 102–109. [CrossRef] [PubMed]

3. Lee, B.G.; Griscom, S.B.; Lee, J.S.; Choi, H.J.; Koh, C.H.; Luoma, S.N.; Fisher, N.S. Influence of dietary uptake and reactive sulfides on metal bioavailability from aquatic sediment. *Science* **2000**, *287*, 282–284. [CrossRef] [PubMed]

4. Fang, H.W.; Chen, M.H.; Chen, Z.H.; Zhao, H.M.; He, G.J. Simulation of sediment particle surface morphology and element distribution by the concept of mathematical sand. *J. Hydro-Environ. Res.* **2014**, *8*, 186–193. [CrossRef]

5. Huang, L.; Fang, H.W.; He, G.J.; Chen, M.H. Phosphorus adsorption on natural sediments with different pH incorporating surface morphology characterization. *Environ. Sci. Pollut. Res.* **2016**, *23*, 18883–18891. [CrossRef] [PubMed]

6. Wang, Y.; Shen, Z.Y.; Niu, J.F.; Liu, R.M. Adsorption of phosphorus on sediments from the Three-Gorges Reservoir (China) and the relation with sediment compositions. *J. Hazard. Mater.* **2009**, *162*, 92–98. [CrossRef]

7. Fang, H.W.; Huang, L.; Wang, J.Y.; He, G.J.; Reible, D. Environmental assessment of heavy metal transport and transformation in the Hangzhou Bay, China. *J. Hazard. Mater.* **2016**, *302*, 447–457. [CrossRef]

8. Huang, S.L.; Wan, Z.H.; Smith, P. Numerical modeling of heavy metal pollutant transport-transformation in fluvial rivers. *J. Hydraul. Res.* **2007**, *45*, 451–461. [CrossRef]

9. Liang, A.; Wang, Y.C.; Guo, H.T.; Bo, L.; Zhang, S.; Bai, Y.L. Assessment of pollution and identification of sources of heavy metals in the sediments of Changshou Lake in a branch of the Three Gorges Reservoir. *Environ. Sci. Pollut. Res.* **2015**, *22*, 16067–16076. [CrossRef]

10. Wang, T.J.; Pan, J.; Liu, X.L. Characterization of heavy metal contamination in the soil and sediment of the Three Gorges Reservoir, China. *J. Environ. Sci. Health A* **2017**, *52*, 201–209. [CrossRef]

11. Feng, C.H.; Zhao, S.; Wang, D.X.; Niu, J.F.; Shen, Z.Y. Sedimentary records of metal speciation in the Yangtze Estuary: Role of hydrological events. *Chemosphere* **2014**, *107*, 415–422. [CrossRef] [PubMed]

12. Gao, L.; Gao, B.; Xu, D.Y.; Peng, W.Q.; Lu, J. Multiple assessments of trace metals in sediments and their response to the water level fluctuation in the Three Gorges Reservoir, China. *Sci. Total Environ.* **2019**, *648*, 197–205. [CrossRef] [PubMed]

13. MEP (Ministry of Environmental Protection of PRC). *The Environmental and Ecological Monitoring Bulletins of the Three Gorges Reservoir Area*; 2016. Available online: http://www.cnemc.cn/jcbg/zjsxgcstyhjjcbg/ (accessed on 5 December 2018).

14. Nilsson, C.; Reidy, C.A.; Dynesius, M.; Carman, R. Fragmentation and flow regulation of the world's large river systems. *Science* **2005**, *308*, 405–408. [CrossRef] [PubMed]

15. Mao, J.Q.; Jiang, D.G.; Dai, H.C. Spatial-temporal hydrodynamic and algal bloom modelling analysis of a reservoir tributary embayment. *J. Hydro-Environ. Res.* **2015**, *9*, 200–215. [CrossRef]

16. Li, Q.F.; Yu, M.X.; Lu, G.B.; Cai, T.; Bai, X.; Xia, Z.Q. Impacts of the Gezhouba and Three Gorges reservoirs on the sediment regime in the Yangtze River, China. *J. Hydrol.* **2011**, *403*, 224–233. [CrossRef]

17. Syvitski, J.P.M.; Vorosmarty, C.J.; Kettner, A.J.; Green, P. Impact of humans on the flux of terrestrial sediment to the global coastal ocean. *Science* **2005**, *308*, 376–380. [CrossRef] [PubMed]

18. Song, Y.X.; Ji, J.F.; Mao, C.P.; Yang, Z.F.; Yuan, X.Y.; Ayoko, G.A.; Frost, R.L. Heavy metal contamination in suspended solids of Changjiang River—Environmental implications. *Geoderma* **2010**, *159*, 286–295. [CrossRef]

19. Huang, L.; Fang, H.W.; Reible, D. Mathematical model for interactions and transport of phosphorus and sediment in the Three Gorges Reservoir. *Water Res.* **2015**, *85*, 393–403. [CrossRef]

20. Huang, L.; Fang, H.W.; Xu, X.Y.; He, G.J.; Zhang, X.S.; Reible, D. Stochastic modeling of phosphorus transport in the Three Gorges Reservoir by incorporating variability associated with the phosphorus partition coefficient. *Sci. Total Environ.* **2017**, *592*, 649–661. [CrossRef]

21. Stone, R. Three Gorges Dam: Into the unknown. *Science* **2008**, *321*, 628–632. [CrossRef]

22. Bing, H.J.; Zhou, J.; Wu, Y.H.; Wang, X.X.; Sun, H.Y.; Li, R. Current state, sources, and potential risk of heavy metals in sediments of Three Gorges Reservoir, China. *Environ. Pollut.* **2016**, *214*, 485–496. [CrossRef] [PubMed]

23. Gao, B.; Zhou, H.D.; Huang, Y.; Wang, Y.C.; Gao, J.J.; Liu, X.B. Characteristics of heavy metals and Pb isotopic composition in sediments collected from the tributaries in Three Gorges Reservoir, China. *Sci. World J.* **2014**. [CrossRef] [PubMed]

24. Gao, B.; Zhou, H.D.; Yang, Y.; Wang, Y.C. Occurrence, distribution, and risk assessment of the metals in sediments and fish from the largest reservoir in China. *RSC Adv.* **2015**, *5*, 60322–60329. [CrossRef]

25. Gao, J.M.; Sun, X.Q.; Jiang, W.C.; Wei, Y.M.; Guo, J.S.; Liu, Y.Y.; Zhang, K. Heavy metals in sediments, soils, and aquatic plants from a secondary anabranch of the three gorges reservoir region, China. *Environ. Sci. Pollut. Res.* **2016**, *23*, 10415–10425. [CrossRef] [PubMed]

26. Wang, L.J.; Tian, Z.B.; Li, H.; Cai, J.; Wang, S.J. Spatial and temporal variations of heavy metal pollution in sediments of Daning River under the scheduling of Three Gorges Reservoir. In Proceedings of the International Forum on Energy, Environment and Sustainable Development (IFEESD), Shenzhen, China, 16–17 April 2016.

27. Wei, X.; Han, L.; Gao, B.; Zhou, H.; Lu, J.; Wan, X. Distribution, bioavailability, and potential risk assessment of the metals in tributary sediments of Three Gorges Reservoir: The impact of water impoundment. *Ecol. Indic.* **2016**, *61*, 667–675. [CrossRef]

28. Zhao, X.; Gao, B.; Xu, D.; Gao, L.; Yin, S. Heavy metal pollution in sediments of the largest reservoir (Three Gorges Reservoir) in China: A review. *Environ. Sci. Pollut. Res.* **2017**, *24*, 20844–20858. [CrossRef]

29. Li, X.; Wei, J.H.; Li, T.J.; Wang, G.Q.; Yeh, W.W.-G. A parallel dynamic programming algorithm for multi-reservoir system optimization. *Adv. Water Resour.* **2014**, *67*, 1–15. [CrossRef]

30. CWRC (Changjiang Water Resource Committee). *Bulletin of the Yangtze River Sediment*; Changjiang Press: Wuhan, China, 2015.

31. Liu, Y.; Liu, H.C.; Li, X.H. Simultaneous and precise determination of 40 trace elements in rock samples using ICP-MS. *Geochimica* **1996**, *25*, 552–558.

32. Page, A.L. *Methods of Soil Analysis, Part 2: Chemical and Microbiological Properties (Agronomy)*; American Society of Agronomy and Soil Science Society of America: Madison, WI, USA, 1982.

33. Fu, J.; Sheng, S.; Wen, T.; Zhang, Z.M.; Wang, Q.; Hu, Q.X.; Li, Q.S.; An, S.Q.; Zhu, H.L. Polycyclic aromatic hydrocarbons in surface sediments of the Jialu River. *Ecotoxicology* **2011**, *20*, 940–950. [CrossRef]

34. Clesceri, L.S.; Greenberg, A.E.; Eaton, A.D. *Standard Methods for the Examination of Water and Wastewater, American Public Health Association*; American Water Works Association and Water Environment Federation: Washington, DC, USA, 1998.

35. Håkanson, L. An ecological risk index for aquatic pollution control—A sedimentological approach. *Water Res.* **1980**, *14*, 975–1001. [CrossRef]

36. Chi, Q.; Yan, M. *Handbook of Elemental Abundance for Applied Geochemistry*; Geological Publishing House: Beijing, China, 2007.

37. Shepard, F.P. Nomenclature based on sand-silt-clay ratios. *J. Sediment. Petrol.* **1954**, *24*, 151–158.

38. Fang, T.; Xu, X.Q. Establishment of sediment quality criteria for metals in water of the Yangtze River using equilibrium partitioning approach. *Resour. Environ. Yangtze Basin* **2007**, *16*, 525–531.

39. Tang, J. Study on the Regularity of Move, Enrichment, and Translation of Cadmium and Other Heavy Metals in the District of the Three Gorges Reservoir. Doctoral Dissertation, Chengdu University of Technology, Chengdu, China, 2005.

40. Xu, X.Q.; Deng, G.Q.; Hui, J.Y.; Zhang, X.H.; Qiu, C.Q. The pollution characteristics of heavy metal in sediments in Three Gorges Reservoir. *Acta Hydrobiol. Sin.* **1999**, *23*, 1–9.

41. CEPA (Chinese Environmental Protection Administration). *Chinese Environmental Quality Standard for Soils (GB 15618-1995)*; 1995. Available online: http://kjs.mee.gov.cn/hjbhbz/bzwb/trhj/trhjzlbz/199603/t19960301_82028.shtml (accessed on 5 December 2018).

42. Wang, X.X.; Bing, H.J.; Wu, Y.H.; Zhou, J.; Sun, H.Y. Distribution and potential eco-risk of chromium and nickel in sediments after impoundment of Three Gorges Reservoir, China. *Hum. Ecol. Risk Assess.* **2017**, *23*, 172–185. [CrossRef]

43. Wang, J.K.; Gao, B.; Zhou, H.D.; Lu, J.; Wang, Y.C.; Yin, S.H.; Hao, H.; Yuan, H. Heavy metal pollution and its potential ecological risk of the sediment in Three Gorges Reservoir during its impounding period. *Environ. Sci.* **2012**, *33*, 1693–1699.

44. Zhang, X.M.; Liu, Y.Y.; Guo, J.P.; Li, Y. Investigation and characterization of heavy metal pollution in sediment of Chongqing urban section of the Three Gorges Reservoir. *J. Chongqing Technol. Bus. Univ.* **2010**, *27*, 176–180. (In Chinese)

45. Chen, H.Y.; Teng, Y.G.; Li, J.; Wu, J.; Wang, J.S. Source apportionment of trace metals in river sediments: A comparison of three methods. *Environ. Pollut.* **2016**, *211*, 28–37. [CrossRef]

46. Feng, L.; Li, C.M.; Hu, B.Q.; Zhang, Y.; Huang, J.S.; Zhang, J. Analysis of pollution characteristics of surface sediments in Three Gorges Reservoir after normal impoundment. *Res. Environ. Sci.* **2016**, *29*, 353–359. (In Chinese)

47. Liu, M.X.; Yang, Y.Y.; Yun, X.Y.; Zhang, M.M.; Li, Q.X.; Wang, J. Distribution and ecological assessment of heavy metals in surface sediments of the East Lake, China. *Ecotoxicology* **2014**, *23*, 92–101. [CrossRef]

48. Ren, D.Y.; Zhao, F.H.; Wang, Y.Q.; Yang, S.J. Distributions of minor and trace elements in Chinese coals. *Int. J. Coal Geol.* **1999**, *40*, 109–118. [CrossRef]

49. Liu, Y.Z.; Xiao, T.F.; Ning, Z.P.; Li, H.J.; Tang, J.; Zhou, G.Z. High cadmium concentration in soil in the Three Gorges region: Geogenic source and potential bioavailability. *Appl. Geochem.* **2013**, *37*, 149–156. [CrossRef]

50. Zhang, C.; Yu, Z.G.; Zeng, G.M.; Jiang, M.; Yang, Z.Z.; Cui, F.; Zhu, M.Y.; Shen, L.Q.; Hu, L. Effects of sediment geochemical properties on heavy metal bioavailability. *Environ. Int.* **2014**, *73*, 270–281. [CrossRef] [PubMed]

51. Xiao, S.B.; Liu, D.F.; Wang, Y.C.; Gao, B.; Wang, L.; Duan, Y.J. Characteristics of heavy metal pollution in sediments at the Xiangxi Bay of Three Gorges Reservoir. *Resour. Environ. Yangtze Basin* **2011**, *20*, 983–989. (In Chinese)

52. Singh, K.P.; Mohan, D.; Singh, V.K.; Malik, A. Studies on distribution and fractionation of heavy metals in Gomati river sediments d a tributary of the Ganges, India. *J. Hydrol.* **2005**, *312*, 14–27. [CrossRef]

water

MDPI

Article

Experimental Investigations of Interactions between Sand Wave Movements, Flow Structure, and Individual Aquatic Plants in Natural Rivers: A Case Study of *Potamogeton Pectinatus* L.

Łukasz Przyborowski [1], Anna Maria Łoboda [1] and Robert Józef Bialik [2,*]

[1] Institute of Geophysics, Polish Academy of Sciences, Księcia Janusza 64, 01-452 Warsaw, Poland; lprzyborowski@igf.edu.pl (Ł.P.); aloboda@igf.edu.pl (A.M.Ł.)

[2] Institute of Biochemistry and Biophysics, Polish Academy of Sciences, Pawińskiego 5a, 02-106 Warsaw, Poland

* Correspondence: rbialik@ibb.waw.pl; Tel.: +48-22-592-5796

Received: 5 August 2018; Accepted: 28 August 2018; Published: 30 August 2018

Abstract: Long-duration measurements were performed in two sandy bed rivers, and three-dimensional (3D) flow velocity and bottom elevation changes were measured in a vegetated area and in a clear region of a river. Detailed flow velocity profiles downstream and upstream of a single specimen of *Potamogeton pectinatus* L. were obtained and the bed morphology was assessed. *Potamogeton* plants gathered from each river were subjected to tensile and bending tests. The results show that the existence of the plants was influenced by both bottom and flow conditions, as the plants were located where water velocity was lower by 12% to 16% in comparison to clear region. The characteristics of the flow and sand forms depended on the cross-sectional arrangement of the river, e.g., dunes were approximately four times higher in the middle of the river than in vegetated regions near the bank. Furthermore, the studied hydrophytes were too sparse to affect water flow and had no discernible impact on the sand forms' movements. The turbulent kinetic energy downstream of a single plant was reduced by approximately 25%. Additionally, the plants' biomechanical characteristics and morphology were found to have adjusted to match the river conditions.

Keywords: aquatic plants; flow velocity measurements; river morphology; acoustic Doppler velocimeter; natural sandy bed river; sand waves; turbulent kinetic energy; aquatic plant biomechanics

---

1. Introduction

Aquatic plants grow in various configurations in rivers, affecting a variety of biological, chemical, and physical processes [1]. They influence both river morphodynamics and flow hydrodynamics by trapping sediments, preventing erosion, affecting turbulent flow fields, and contributing to overall bed roughness, as reported in numerous studies, e.g., [2–8]. One of the most important conclusions of these studies was that the understanding of flow–biota–sediment interactions requires an interdisciplinary approach. O'Hare et al. [9] emphasized that linking plant traits with physical modeling is an example of such multidisciplinary research, where ecological variability is a valid factor. Furthermore, Reid and Thoms [10] highlighted the need to obtain three-dimensional (3D) velocity and turbulence measurements to properly describe river habitats.

In vegetated rivers, sediment transport and channel morphodynamics are affected by vegetation [1,11]. In a vegetated location, the bed load is expected to diffuse transversely into the vegetation and deposit in the lee side of the patches [12], with increased erosion at the lateral sides of the patches, constituting the so-called scouring effect [13]. Suspended sediment is prone to being deposited in the wake of vegetation [14]. Bouma et al. [15] observed high turbulence and erosion at the leading edges and sides of patches of epibenthic structures such as *Spartina anglica* tussocks. This intertidal vegetation strongly reduced the water velocity, resulting in sedimentation farther downstream. Moreover, Rominger et al. [13] described how vegetative drag diverges and accelerates the flow, potentially causing erosion at the patch edge, which is similar to the scour patterns observed in the field around individual flow obstructions such as bridge piers. In addition, Schnauder and Sukhodolov [16] investigated the seasonal patch effect on sediment transport and concluded that although plant patches affected the transverse flow profiles near the edges, pool scouring did not occur; however, sediment was observed to accumulate in the recirculation zones. Additional data were reported by Cavedon [17], who, based on laboratory experiments, concluded that for a sufficiently high density of stems, the length of bed forms was not influenced by vegetation density, but only by the distance between plants. Velocity profiles in vegetated channels change throughout the year due to plant growth and senescence, therefore shear velocity influencing beds also varies [18]. Our investigation revealed that the height of the forms is reduced first in the vegetated parts of a river. Then it depends on vegetation density and is strongly coupled to the characteristics of the flow field and sediment properties. However, this phenomenon requires further analysis, particularly in natural field conditions.

From a broader perspective, the sedimentation rates in a river are time-dependent, thus natural aquatic habitat characteristics always depend on flow, biota, and morphodynamic feedback over a certain time scale [5,19]. Consequently, the species living in these habitats are called ecosystem engineers because of their control over flow velocity and sediment deposition [20]. To find patterns in the changing and complex ecosystem of a vegetated river, most researchers have focused on flume experiments (e.g., [8,21–23]), and only a few have conducted similar measurements in natural rivers (e.g., [24,25]). Such experiments are typically conducted using an acoustic Doppler velocimeter (e.g., [24,26–29]), which allows measurement of the velocity field at discrete points over bed forms. Although this device was built for laboratory purposes, it is usable in the field (e.g., [30]), albeit with certain limitations in its use and in interpreting the acquired data [31,32]. One can characterize changes in bed structure using either a discrete or continuous approach [33,34]. In the present study, the former approach was used to characterize bed forms in terms of height, length, and celerity, using the echo-sounding ability of current profiler and velocity profiler.

The main goal of this study was to evaluate changes and differences in sand wave movements and flow in vegetated and unvegetated parts of a channel cross-section with regard to sediment composition, flow dynamics, and bed morphology in two natural lowland rivers. This was achieved by recording these changes with long-duration point measurements of 3D velocity and bed elevation, and by scanning bed morphology in chosen parts of rivers with aquatic plants, which gave sufficient quantitative and qualitative information about the investigated subject. Moreover, the biomechanical features of a chosen aquatic macrophyte were studied to investigate how these characteristics differ between plants collected from two distinct river habitats. Due to the challenges inherent in the task of studying multiple aspects of a river ecosystem simultaneously, this case study also represents an attempt to identify the needs and limitations in a multidisciplinary investigation of flow–biota–sediment interactions in field conditions.

2. Materials and Methods

2.1. Equipment

Measurement of the instantaneous three-dimensional (3D) velocity field and recording of the bottom elevation were performed using a Vectrino Profiler (VP) (revision 2779/1.32, Nortek AS, Rud, Norway). The following VP setup parameters were chosen: Recording frequency of 25 Hz, minimum ping algorithm with high power level, and velocity range up to 0.1 m·s^{-1} above the maximum observed longitudinal velocity during a trial. VPs were mounted on a steel platform. In addition, the bathymetry and velocity spatial distribution of the rivers were obtained using an acoustic Doppler current profiler (ADCP), RiverSurveyor S5 model (SonTek, San Diego, CA, USA). The grain-size distribution was measured using a laboratory shaker with a standard sieve set (10 mm, 2 mm, 1 mm, 0.50 mm, 0.25 mm, 0.10 mm, and 0.071 mm mesh). Before sieving, the sediment sample was dried at 105 °C.

2.2. Study Sites

The first experiment was conducted on the Jeziorka River on 4 July 2017. This river is located south of Warsaw (52°04′55.2″ N, 21°04′02.4″ E). The vegetation in this location covered no more than 20% of the channel cross-section. The water temperature (~15 °C) and other basic characteristics of the river, as well as bed morphology in regard to actual water depth, were measured using the ADCP (Table 1). Bed sediment samples from 3 random locations across the river were taken to conduct granulometry and were found to consist of fine, moderately sorted sand with $D_{16} = 0.11$ mm, $D_{50} = 0.20$ mm, and $D_{84} = 0.39$ mm, and an inclusive graphic standard deviation equal to 0.848 (Figure 1).

The platform in the first part of the experiment, where the velocities in the whole water depth were measured, was placed parallel to the flow, above a single specimen of *P. pectinatus*. Measurements were conducted at a point located 20 cm upstream of the plant and at points 25 and 50 cm downstream, in the wake of the plant, in which flow is supposedly affected by it. Recording of velocity in each point of the velocity profile lasted 3 min. The hydrophyte was then collected for further biomechanical testing.

In the second part of the experiment, the platform was placed perpendicular to the flow. The first VP was placed 20 cm in front of another specimen of *P. pectinatus* in a 55 cm deep vegetated region in the river adjacent to the right bank, where the majority of the plants grew (Figure 2). The second device was placed near the middle of the river, where the bed was bare. With both probes positioned 15 cm above the bottom level, the velocities and bottom elevation changes were measured for 1.5 h. A survey of the river bottom adjacent to the right bank revealed many small specimens of *P. pectinatus* and a substantial amount of debris, mostly in the form of dead wood in the vegetated region.

The second experiment was conducted on the Świder River (52°07′59.2″ N, 21°15′41.5″ E) on 1 October 2017. The riverbed in this location was covered with sand and was almost clear of plants, which grew only along the left bank. The water temperature (~11.6 °C) and other basic characteristics of the river, as well as bed morphology regarding actual water depth, were measured using ADCP (Table 1). In addition to *P. pectinatus*, two other species (*Myriophyllum spicatum* L. and *Potamogeton crispus* L.) were also found growing downstream at the left bank. The platform was oriented perpendicular to the flow, 1 m from the left bank (Figure 3). The first velocimeter was positioned 10 cm upstream along a 0.6 m long and 1 m wide strip where *P. pectinatus* specimens were located. The second velocimeter was placed in a clear region 3 m from the first one. The bed elevation and velocity recording lasted 3 h.

Granulometry revealed that the bed sediment in the Świder River consisted of medium poorly sorted sand with $D_{16} = 0.14$ mm, $D_{50} = 0.24$ mm, and $D_{84} = 0.47$ mm, and an inclusive graphic standard deviation equal to 1.02 (Figure 1).

Table 1. Basic hydraulic statistics of the studied rivers in the experimental locations.

	Mean U (m·s^{-1})	Discharge Q (m^3·s^{-1})	Mean H (m)	Width (m)	Reynolds Number (U·H·v^{-1})	Froude Number (U·(H·g)$^{-0.5}$)
Jeziorka R.	0.30	2.01	0.6	11	1.58×10^5	0.123
Świder R.	0.51	4.26	0.46	18	1.88×10^5	0.24

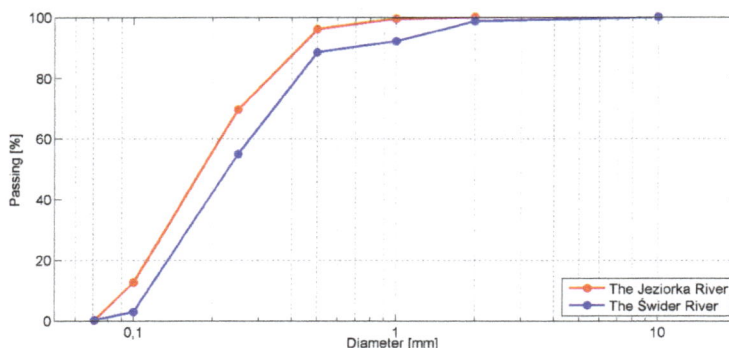

Figure 1. Granulometric distribution curves of bed sediment collected from the Jeziorka and Świder Rivers.

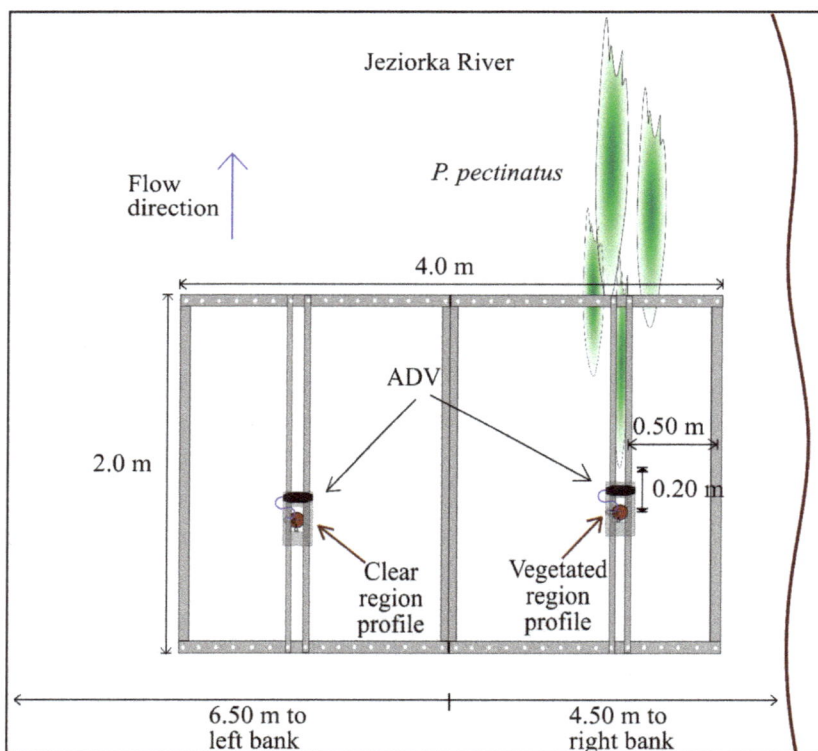

Figure 2. Arrangement of the platform with Vectrino Profilers and individuals of *P. pectinatus* during the experiment in the Jeziorka River (second part).

2.3. Plant Characterization

Hydrodynamic measurements were conducted near the same aquatic plant species, *P. pectinatus*, which is a slender, submerged macrophyte [35]. Each field experiment was followed by laboratory measurements of biomechanical traits.

In the Jeziorka River, the chosen individual plant had a length of 1 m and floated almost horizontally approximately 10 to 20 cm above the bottom. Other macrophytes in the vegetated region of the channel grew in a dense patch 5 m in length, and their stems floated higher above the bottom than the tested hydrophyte. In the Świder River, individuals of *P. pectinatus* were much shorter, with an average length of 35 cm. The investigated plants grew in a different pattern than the macrophytes in the Jeziorka River; instead of one slender and long patch, the plants were spread across a 1.5 m wide and 0.6 m long area (Figure 3), with stems floating at a maximum of 5 cm above the riverbed.

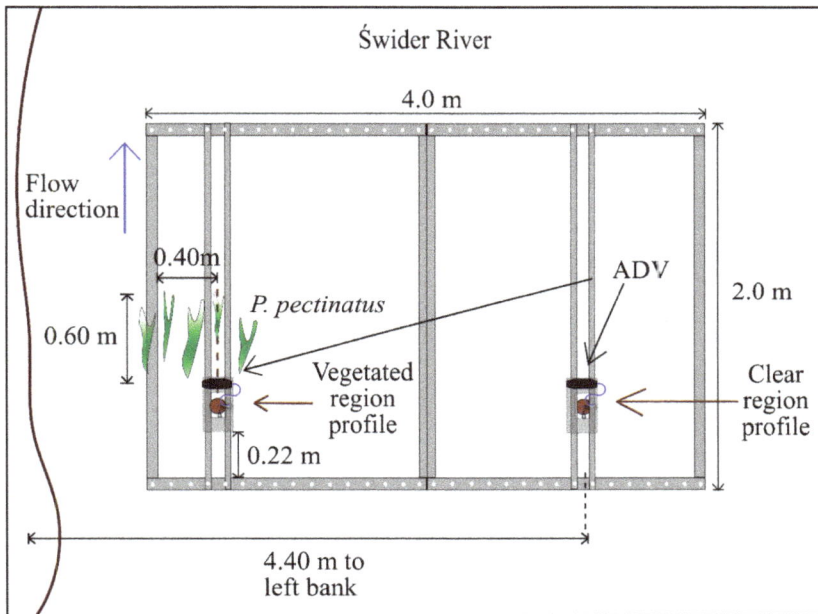

Figure 3. Arrangement of the platform with Vectrino Profilers and individuals of *P. pectinatus* during the experiment in the Świder River.

2.4. Measurements of Biomechanical Traits

Biomechanical tests (3-point bending and tension tests) were conducted using a Tinius Olsen Bench Top Testing Machine, 5ST Model (Tinius Olsen, Redhill, UK), and the data were recorded using Horizon software. The plants were transported to the laboratory and kept in a 112 L tank (more details are available in [36]). After a plant was removed from the aquarium, the stem was cut into 7 cm pieces. Next, the diameter of each sample was measured using a microscope. In the tension tests, short strips of sandpaper were glued to the ends of the samples to prevent them from slipping out of the machine clamps [36]. The measurements were conducted in submerged conditions [36]. In addition, to minimize the time interval between removing a plant from the aquarium and conducting the test, no more than 10 samples were prepared at the same time.

The investigated parameters in the 3-point bending test included maximum force, maximum stress, maximum deflection, flexural strain, flexural rigidity, and flexural modulus, and the tension parameters included breaking force, stress and strain, and Young's modulus. All parameters were calculated following formulas from Łoboda et al. [36], based on Niklas [37] and ASTM [38].

2.5. Data Processing

The VP was used to record velocities in a profile placed 11 cm above the bottom at the start of the measurement in the Jeziorka River and 15 cm in the Świder River. The depth was chosen to perform measurements in a water volume as close to the bottom as possible, but without danger of burying the probe. In case of profiles upstream and downstream of *P. pectinatus* in the Jeziorka River, each point in the profile represented a 3-min recording, and the vertical distance between the following points was 3 cm. The collected datasets were then processed and filtered before further analysis. Parts of code included in the MITT open source MATLAB algorithms [39] were used for velocity data processing. The time-dependent analysis of the recorded velocity data revealed significant drops in data quality caused by noise that affected certain VP beams. Thus, a significant part of velocity data records was discarded during the second experiment.

To exclude spikes in the velocity records from the VP, a 3-stage method was used. First, from each measurement point containing 30 cells within it, one cell within the "sweet spot" was identified with the best signal-to-noise ratio (SNR) value [32,40,41] and subjected to further processing. Second, points in the recording with correlation below 70% and SNR values below 15 dB were discarded and replaced using linear interpolation. The last step included using the modified 3D phase-space thresholding method filter to remove the spikes [42–44].

After the filtering, the following flow characteristics were calculated: Mean velocity, standard deviation of velocity, turbulent kinetic energy, turbulence intensity, and Reynolds stress (Tables 3 and 4). The turbulent kinetic energy (TKE) was calculated as:

$$\text{TKE} = \frac{1}{2}\left(\overline{(U')^2} + \overline{(V')^2} + \overline{(W')^2}\right)$$

where U', V', and W' are velocity fluctuations in the longitudinal, lateral, and vertical directions, respectively, and the overbars denote mean values. Normal Reynolds stresses, turbulence intensities, and TKE were calculated as noise free, which was possible due to redundant data of 2 independent vertical velocities recorded by VP [45,46]. As the first vertical velocity was obtained from the same pair of beams as longitudinal velocity, it was used for all relevant calculations [41].

The power spectral densities were calculated using scripts implemented in MATLAB, i.e., the Welch [47] method with discrete Fourier transform points equal to 512, with 50% overlap and a Hamming window function [41].

The characteristics of the bed forms were calculated from the data acquired by the current profiler, where the lengths of forms were given as the height from the local maximum crest to the local minimum trough [48]. The mean celerity of the bed forms was estimated from the current profiler and VP records and was calculated using the equation proposed by Nikora et al. [33]:

$$C = \frac{0.66\overline{U}Fr^{2.9}}{\frac{g\lambda}{\overline{u}^2}},$$

where $\overline{U}$ is mean velocity, Fr is the Froude number, g is gravitational acceleration, and λ is the length of a bed form.

3. Results

3.1. Bed Elevation Long-Duration Behavior and Bed Morphology

The bed morphology revealed by a river cross-section (Figure 4) showed uneven elevations near the left bank of the river and a relatively flat bottom through the middle part up to the deepest section near the right bank, where the majority of the *P. pectinatus* specimens grew. The results from the experiment conducted in the Jeziorka River showed fluctuations of the bed elevation of less than 0.5 cm in the vegetated area and up to 2 cm in the clear area (Figure 5).

The experiment conducted in the Świder River showed ripples 1.0 to 2.5 cm high in the vegetated area (Figure 6). In the clear region, no distinct ripples were observed. In the clear region, a sand dune 10 cm high was observed (Figure 6). The occurrence of smaller ripples was then observed on the upstream side of that sand form, which were clearly distinguishable as a second type of wave form [33,49]. The streamwise depth elevation profiles revealed easily distinguishable deep and shallow areas close to the bank (Figure 4, profile I in Figure 7). Moving away from the bank, profile II, which was at the same distance from the bank as the long-duration measurement point in the vegetated area, showed continuous changes in elevation approximately 3 cm high. Profile IV, which was at the same distance from bank as the long-duration measurement point in the clear area, showed a series of dunes 10 cm high. Both profiles II and IV were thus consistent with time-dependent changes in the vegetated and clear regions (Figure 6). The measured properties of the sand forms showed that height and length increased with increasing distance from the channel's left bank, while dune celerity decreased (Table 2, Figure 7).

Table 2. Statistics of sand waves in the considered profiles in the Świder River. L denotes length of a dune; H denotes height of the dune; SD denotes standard deviation.

Location	Mean L (cm)	Mean H (cm)	SD L (cm)	SD H (cm)	Mean Time τ (s)	Celerity (ms^{-1} × 10^3)	Theoretical Celerity (ms^{-1} × 10^3)
Profile II	127.1	3.2	36.0	2.2	1338	0.949	0.112
Profile III	257.5	12.8	67.9	5.2	–	–	–
Profile IV	345.8	13.2	138.5	4.0	10.400	0.332	0.041

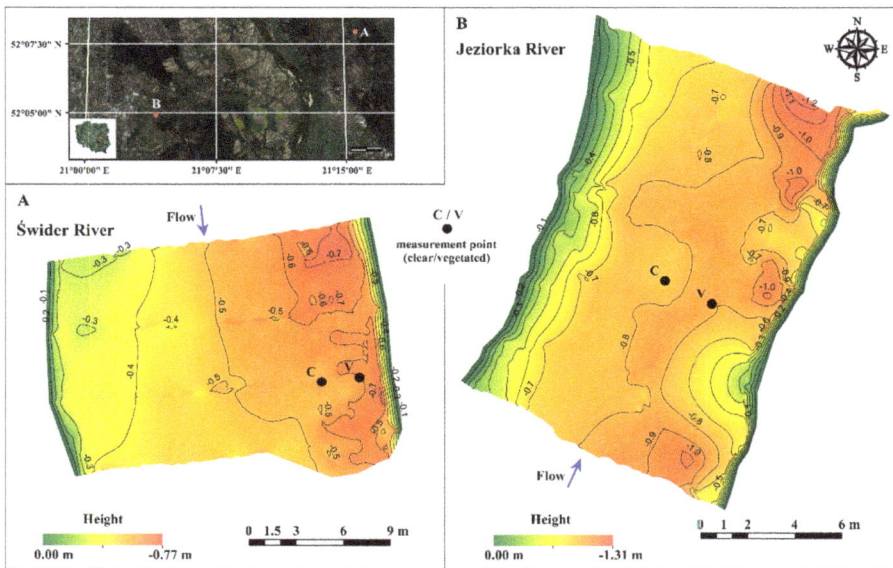

Figure 4. Bed morphology of studied sites in the (**A**) Świder River and (**B**) Jeziorka River.

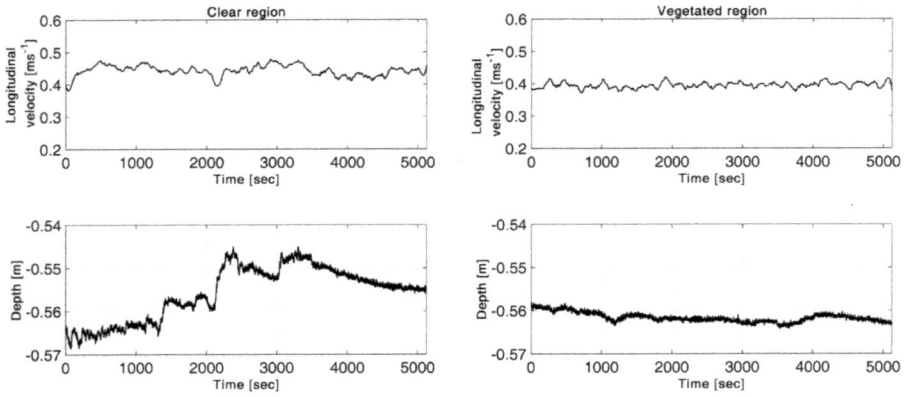

Figure 5. Smoothed longitudinal velocity and depth changes in the Jeziorka River at one point in clear, and one in vegetated region.

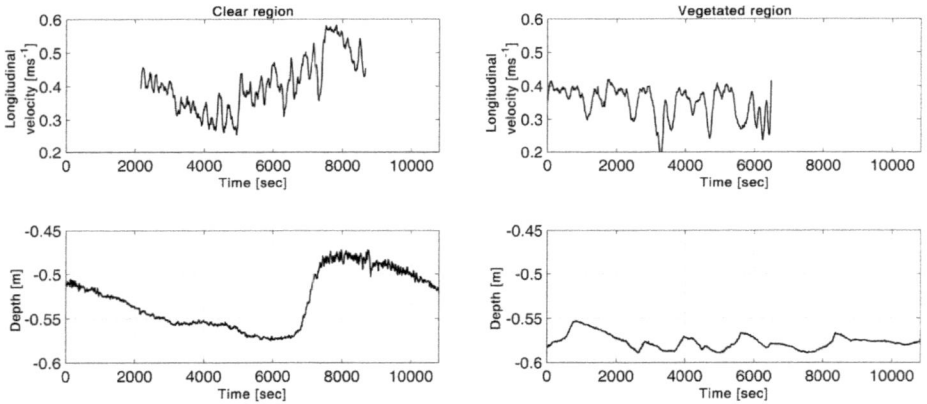

Figure 6. Smoothed longitudinal velocity and depth changes in the Świder River at one point in clear, and one in vegetated region. Parts of the velocity recording were removed due to bad signal characteristics.

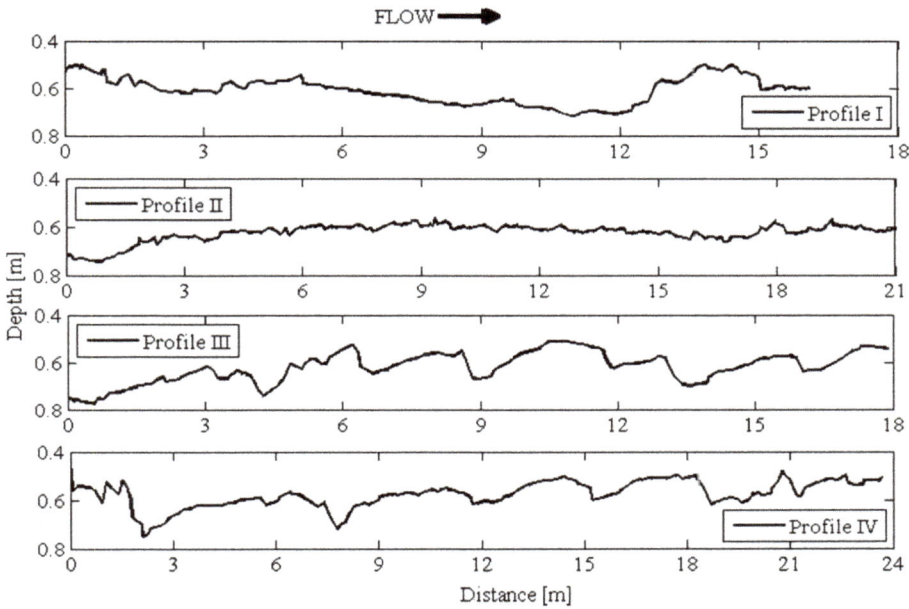

Figure 7. Bed elevation profiles made in the stream-wise direction in the Świder River. Please note that profile numbers increase with distance to the left bank.

3.2. Velocity Long-Duration Behavior

The long-duration velocity measurements in two distinct river regions were quasi-stationary, given various phases of sand waves moving beneath, and represent only quantitative differences between those regions and the rivers themselves (Tables 3 and 4). The measured mean longitudinal velocity was higher in the clear flow path in the Jeziorka and Świder rivers by 12% and 16%, respectively, with mean squared error two times higher in the latter (Table 3). The lateral and vertical velocity fluctuations were also higher in the clear area in the Świder River, while in the Jeziorka River there was no difference between regions (Table 3). The normalized turbulence intensities in the Świder showed increased mixing in longitudinal and vertical directions (higher in clear region than in vegetated region), in contrast to the Jeziorka River, where lateral mixing increased with no differences between regions (Table 3). The turbulent kinetic energy was the highest in the clear region in the Świder River, three times higher than in the other three regions. The normal stresses in the clear region of the Świder River were larger than those in the vegetated area (Table 4), particularly the shear stress ($U'W'$), which showed typical values as reported by Sukhodolov et al. [27] above the bed with a sandbar of similar height (albeit immobile). The ratio of normal transverse and vertical stresses was close to 1.65 for the Jeziorka River, which is consistent with the coefficient for open channel flows found by Nezu and Nakagawa [50]. The fluctuating longitudinal velocity power spectral density (PSD) plot (Figure 8) showed the occurrence of a $-5/3$ Kolmogorov scaling range at higher frequencies, and a -1 scaling range at lower frequencies, which are attributed to the inertial and production subranges of turbulence, respectively [51]. In the Jeziorka River, there were no apparent differences in the PSD between the clear and vegetated regions, in contrast to the Świder River, where the PSD was flatter for the region with plants.

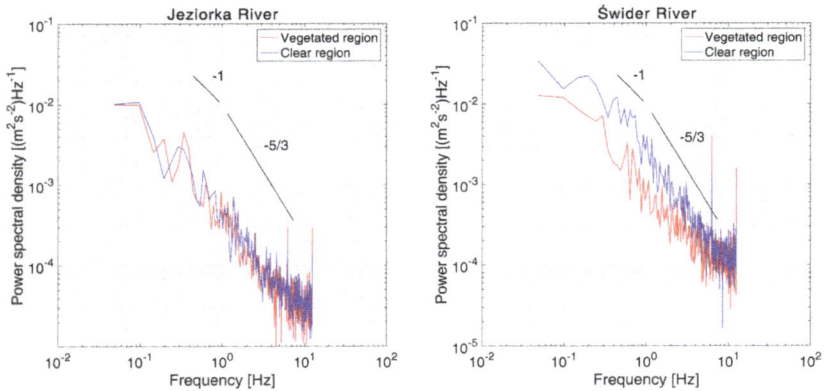

Figure 8. Power spectrum of longitudinal velocities calculated from the filtered signal of long-duration measurements.

Table 3. Mean velocity, turbulence intensity, and turbulent kinetic energy of long-time measurements.

Location		Mean Velocity ± SD (cm·s⁻¹)			Turbulence Intensity (Normalized)			Turbulent Kinetic Energy (m²·s⁻²)
		u	v	w	$\frac{\sqrt{\overline{u'^2}}}{\overline{u}}$	$\frac{\sqrt{\overline{v'^2}}}{\overline{u}}$	$\frac{\sqrt{\overline{w'^2}}}{\overline{u}}$	
Świder R.	Vegetated	34.95 ± 9.61	−4.52 ± 5.90	0.52 ± 4.25	0.2115	0.0437	0.1000	0.0032
	Clear	40.56 ± 15.14	−5.81 ± 10.05	−5.56 ± 8.41	0.2851	0.0628	0.1972	0.0102
Jeziorka R.	Vegetated	39.46 ± 6.18	8.05 ± 4.90	−0.62 ± 3.53	0.1521	0.1186	0.0885	0.0035
	Clear	44.25 ± 6.21	3.28 ± 4.81	−2.95 ± 3.66	0.1345	0.1011	0.0816	0.0034

Table 4. Mean velocity, turbulence intensity, and turbulent kinetic energy of long-time measurements.

Location		Reynolds Stresses (kg·m⁻¹·s⁻²)					
		Normal Stresses			Tangential Stresses		
		$-\rho\overline{u'u'}$	$-\rho\overline{v'v'}$	$-\rho\overline{w'w'}$	$-\rho\overline{u'v'}$	$-\rho\overline{u'w'}$	$-\rho\overline{v'w'}$
Świder R.	Vegetated	−5.46	0.23	−1.22	−0.27	0.91	0.28
	Clear	−13.37	−0.65	−6.40	1.72	5.68	−0.93
Jeziorka R.	Vegetated	−3.60	−2.19	−1.22	0.67	0.75	−0.03
	Clear	−3.54	−2.00	−1.30	−0.04	0.81	−0.13

3.3. Velocity Profiles in the Proximity of P. pectinatus

The mean longitudinal velocities upstream of *P. pectinatus* were distributed in the expected logarithmic profile, with a maximum value of 0.58 m·s⁻¹ (Figure 9). Two downstream profiles were similar in each velocity direction, but with increased deviations below Z/H = 0.3, especially for the profile 0.25 m behind the plant. The profile upstream had increased mean lateral velocity and vertical velocity oscilating near 0 m·s⁻¹, while downstream of the plant the velocity tended to be directed toward the bottom. The turbulent kinetic energy was elevated at a height of Z/H = 0.1 upstream of the plant, but downstream it was lower at that height (Figure 9).

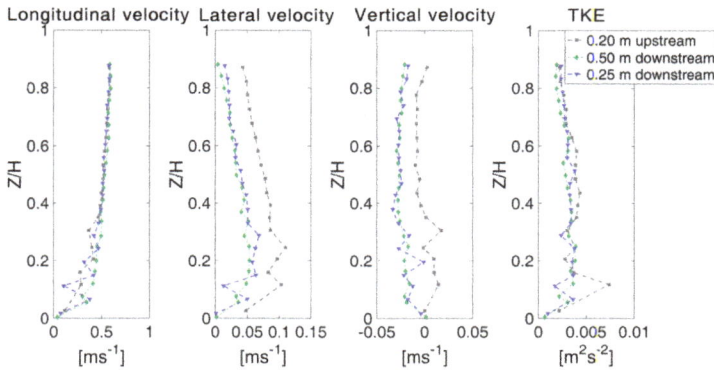

Figure 9. Profiles from the Jeziorka River upstream and downstream of *P. pectinatus*.

3.4. Biomechanical Measurements

P. pectinatus samples from the Jeziorka River were characterized by thinner stems than those collected from the Świder River. The mean diameters in the three-point bending tests were 0.99 and 1.14 mm for hydrophytes from the Jeziorka and the Świder rivers, respectively (Table 5), whereas in the tension tests, these values ranged from 0.98 mm for species from the Jeziorka to 1.33 mm for individuals from the Świder (Table 6). The maximum force, maximum stress, and flexural strain were similar for both locations (Table 5). However, *P. pectinatus* from the Jeziorka River was characterized by lower flexural rigidity and flexural modulus of 7.82 N·mm² and 175.20 MPa, respectively (Table 5). The same species from the Świder River had a stiffness that was approximately twice as high: The flexural rigidity was 16.09 N·mm² (Table 5). The flexural modulus was also much higher, reaching 280.39 MPa.

The tension test results show that the breaking force of *P. pectinatus* from the Jeziorka River was higher, 3.32 N compared to the value associated with the Świder River, 2.34 N (Table 6). The differences in strength were more prominent; the plant from the first experiment had a breaking stress of 4.56 MPa, whereas the value was lower for the second plant, at 1.91 MPa (Table 6). The strain of the studied species from the Jeziorka River was 12.33%, whereas for hydrophytes from the Świder River, the strain was 9.08%. *P. pectinatus* displayed a mean Young's modulus of 104.50 MPa in the plants from the first measurement and 38.17 MPa in the plants from the second (Table 6).

Table 5. Diameter and three-point bending test results for *P. pectinatus*.

Parameter		Jeziorka River		Świder River	
		5 July		1 October	
		20 Samples		22 Samples	
		Mean ± SD	Median	Mean ± SD	Median
Diameter	(mm)	0.99 ± 0.17	0.99	1.14 ± 0.31	1.01
Maximum force	(N)	0.022 ± 0.007	0.020	0.024 ± 0.017	0.020
Maximum stress	(MPa)	0.030 ± 0.013	0.025	0.023 ± 0.010	0.023
Flexural strain	(%)	4.38 ± 1.02	4.27	4.52 ± 1.73	4.33
Max. deflection	(mm)	16.14 ± 1.55	16.13	14.57 ± 2.97	15.62
Sec. m. of area	(mm⁴)	0.057 ± 0.049	0.046	0.131 ± 0.209	0.051
Flexural rigidity	(N·mm²)	7.82 ± 4.22	6.22	16.09 ± 5.27	16.07
Flexural modulus	(MPa)	175.20 ± 91.49	159.80	280.39 ± 188.84	271.75

Table 6. Diameter and tension test results for *P. pectinatus*.

Parameter		Jeziorka River		Świder River	
		5 July		1 October	
		20 Samples		19 Samples	
		Mean ± SD	Median	Mean ± SD	Median
Diameter	(mm)	0.98 ± 0.12	1.00	1.33 ± 0.35	1.28
Breaking force	(N)	3.32 ± 0.63	3.45	2.34 ± 0.77	2.22
Breaking stress	(MPa)	4.56 ± 1.30	4.63	1.91 ± 0.95	1.62
Breaking strain	(%)	12.33 ± 3.42	13.18	9.08 ± 2.17	8.84
Young's modulus	(MPa)	104.50 ± 35.19	104.00	38.17 ± 26.01	35.83

4. Discussion

The typical configuration of aquatic plants found in both studied rivers consisted of a strip of plants growing along the river bank, similar to the configuration observed in the study by Naden et al. [24]. We expect aquatic macrophytes to grow closer to river banks due to the favorable conditions for seeds to be trapped [52]. In the case of *P. pectinatus*, Chambers et al. [53] showed that it thrives in rivers with slower currents; therefore, in the vegetated region, the water current was expected to be lower than in the main stream due to the plant's growth requirements. This theory was borne out by the obtained results, with measured velocity being 12% and 16% lower than that in the clear regions in the Świder and Jeziorka rivers, respectively (Table 3). However, the study of Kemp et al. [54] showed that submerged macrophytes with fine leaves thrive in habitats with a wide range of Froude numbers.

The investigated rivers exhibited similar granulometry (Figure 1), but the Świder River had a Froude number two times greater (Table 1), and it produced different sand wave patterns. In the Jeziorka River, approximately 20% of the width of the studied cross-section was considered the vegetated region, while in the Świder River, the living plants spread to no more than 10% of the channel width. Note that the experiments were conducted in meandering rivers, which is reflected in the cross-sectional structure (Figure 4), and which in turn had a major influence on bed scour and deposition [55,56]. Three regions were distinguished in both rivers: The relatively flat, deepest region with vegetation, a middle region without plants but with sand forms, and a shallow region opposite the deep one (Figure 4), which in the Jeziorka River was also vegetated. This pattern of deep and shallow fragments of a channel, a sequence of pools and riffles, is typical for natural meandering rivers [57–59]. However, the pattern of plant patches in a channel varies not only yearly but also seasonally, changing the bed shear stress over time [18,24]. Furthermore, the occurrence of plants and flattening of sand waves may be connected to the appearance of dead wood debris, which was found in the vegetated regions in both rivers, as this type of debris can trap seeds and organic matter [60]. Submerged vegetation in this case performs a similar role as emergent or riparian vegetation: Trapping fine sediment, building landforms, and ultimately stabilizing the river banks [19,61,62]. This hypothesis is consistent with the occurrence of sandbars, visible in vegetated regions near banks in both rivers (Figure 4). These bars could serve as plant-rooting locations, where vegetation decreases the flow velocity within the patch and increases it around the main body of the plant [63]. The present case study revealed that sparsely rooted plant specimens did not create blockage factor big enough to deflect flow and scouring effect was not observed, which otherwise would protect a plant from being buried by passing sand forms.

The collected records of bed elevation changes show differences between various regions in river cross-section morphology in terms of the discrete characterization method. The Jeziorka River featured completely disturbed sand waves in the vegetated region (Figure 5), while in the Świder River, these waves occurred in the vicinity of *P. pectinatus*; however, these dunes were three times lower than the fully developed wave forms in the clear region (Figure 6). The bed form celerity shows that larger dunes were migrating at slower rate than ripples close to the river bank (Table 2). Large bed forms contain more bed load, therefore a significant part of the sediment must be transported in the clear region of the river through dune movements [64]. The flow in that region was characterized by higher Reynolds stresses due to higher shear caused by larger dunes and higher mean longitudinal velocity, however reduced in the dune trough (Figure 6), which is in line with those of, e.g., Venditti and Bauer [65] and Kostaschuk [66]. In comparison to the study of Sukhodolov et al. [27], mean Reynolds shear stress was 10 times higher in spite of the same mean velocity, however it can be explained by the fact that measured point in the present case study in the Świder River was relatively close to the bottom and a lot of negative velocities were recorded, which was not the case in study of Sukholodov [27].

A spike visible in the power spectral density plot in the clear region of the Świder River at 0.4–0.5 Hz (Figure 8) can be related to eddy-like flow structures that originated from the flow separation zone in the lee side of the dune [50]. However, Venditti and Bauer [65] indicated that eddies could not be attributed to a certain frequency, and what is more, analysis of a long-duration velocity power spectrum can be misleading due to the occurrence of local nonstationarity in the time series (particularly prominent in Figure 6). Furthermore, Singh et al. [67] reported that eddies with a vertical size smaller than the distance from the bottom to the measurement point could not be found in the spectra at all. Nevertheless, the spectrum of the velocity in the vegetated region in the Świder River was flatter in the inertial subrange than in the clear region. Consequently, more energy was conducted into higher frequencies in the clear region, as would be expected in the dissipation wake of a bed form [65]. The elevated Reynolds stress in that clear region (Table 4) was also in agreement with the observation of energy distribution. Obtained results indicate that there was a connection between the height of sand dunes and Reynolds shear stress in the water layer above. However, in the pool region, Reynolds stress distribution might be not correlated with sediment size or flow characteristics [59].

Plants, even of the same species, that grow in similar rivers can be characterized by changes in biomechanical properties that depend on the habitat conditions and the phase of plant growth [68]. *P. pectinatus* from two investigated rivers differed in terms of length, thickness, and biomechanical traits. Hydrophytes from the Świder River were shorter, thicker, and stiffer (Tables 5 and 6), and were more resistant to bending forces than specimens from the Jeziorka River. Nevertheless, they were more durable in the tension tests (Table 6). In other words, the specimens growing in the region where the bed was flat grew longer and had altered characteristics such that they were more prone to bend to withstand higher forces resulting from their length. This behavior may be an example of an avoidance strategy, as described by Puijalon et al. [69]. The occurrence of drag on a simple specimen was visible in the measured velocity profiles (Figure 9), in which a region of disturbed velocities in the plant wake was noted. In the Świder River, the increased resistance to bending by specimens growing out of the riverbed, where the ripples were as high as the plants themselves, may be a natural strategy to withstand unfavorable conditions. Then, the most likely reason for the above-mentioned differences in biomechanical properties of hydrophytes may be connected to their diversity of growth phases [68].

The question arises as to whether the studied plants affect the flow and sediment movements or vice versa. The measurements in the Jeziorka River revealed lower mean longitudinal and lateral velocities in the wake of the single plant, but most outstanding was the lower TKE relative to that of the point upstream (Figure 9). The TKE results were not consistent with those obtained by Biggs et al. [70] with another hydrophyte of similar morphology, *Ranunculus penicillatus* (Dumort.) Bab., for which TKE was elevated. However, that study focused on a dense patch and not on a thin individual plant. In the Jeziorka River, the studied macrophyte was streamlining, which is a natural behavior for flexible plants in high-flow conditions [71,72]. Moreover, Siniscalchi and Nikora [72] showed that single hydrophytes exhibited flapping-like motions to minimize drag force, indicating their susceptibility to turbulent eddies [63]. With the frontal area of *P. pectinatus* being approximately four times smaller than that of the patch in Biggs's experiment, the effect on flow in the plant's wake was barely discernible (Figure 9). The studied individuals were too small to show any effect on the bottom morphology; however, there were denser patches downstream (Figure 2), which were followed by a shallower area (Figure 4). In the Świder River, plants grew too small and sparsely to have a significant impact on the velocity profile or bed shear stress (canopies of a/h << 0.1, where *a* is diameter and *h* is water height; see [71]) or even on the sand forms; therefore, plants were periodically covered by migrating dunes, as we observed in our measurements. This result is consistent with a theory by Green [63], who suggested that certain hydrophytes in regions close to the bank and the riverbed do not contribute significantly to resistance due to drag of the boundary layer itself. The present case study investigated such species with even lower shoot density and more sharp-ended stems than *R. penicillatus*, which have practically no blockage effect. This can be a hint, which species are likely to positively contribute to biodiversity of river but not significantly affect flow resistance. Furthermore, Chen and Kao [73] showed that flow through sparse vegetation was similar to flow that occurs in a solid-boundary channel.

Studies performed in situ by Naden et al. [24] and Sukhodolova and Sukhodolov [74] represent a goal for how studies of flow–biota–sediment interactions should be conducted. The present investigation was smaller in scale and details, as the requirement of maintaining stationarity of the measurement and revision of local conditions forced us to skip some of the observations. In spite of this, the findings of the presented study show agreement with general theories on this subject and add scarcely available data to it. Therefore, the next step would be to perform subsequent measurements in the next summer season in the same spots to observe changes in bed morphology and vegetation cover. Additionally, the creation of a denser grid of velocity profiles around *P. pectinatus* patches, including big canopies present in the Jeziorka River, would be preferable in order to find how these plants' assembles affect flow field.

5. Conclusions

We present the following observations with regard to the plant occurrence, sediment movements, plant traits, and effects of hydrophytes on flow and sand forms:

- The aquatic plant *P. pectinatus* grew in a larger quantity in the river with the lower Reynolds number, tending to be located in a pool section of a channel, where smaller ripples occurred. This finding suggests an emerging pattern in which flow conditions and bed morphology are connected with the occurrence of the studied hydrophyte.
- The bed forms measured in the Świder River tended to exhibit increased height and length with increased distance from the vegetated region close to the bank. These forms were also more developed than those in the Jeziorka River, which is characterized by finer sediment and faster but steadier flow, with lower turbulent kinetic energy and a lower Froude number.
- Major differences were identified in the plants' morphology and biomechanics: Older individuals from the faster-flowing Świder River were thicker, shorter, and stiffer than their younger counterparts from the Jeziorka River, which were more prone to bending. These traits may have resulted from the aquatic macrophytes' adjustment to the habitat conditions.
- Strips of short individual *P. pectinatus* plants within a mobile bed in the Świder River did not seem to detectably affect passing sand ripples. Turbulence statistics suggest a much steadier flow than that in the clear area.
- The single hydrophyte from the Jeziorka River did not affect the velocities measured downstream; only elevated TKE was visible in the wake of the plant.
- The mean velocity in front of the plants was approximately 12% and 16% lower than in the clear regions in the Świder River and Jeziorka River, respectively. Therefore, bed conditions such as small sandbars and wood debris trapping organic material could have made major contributions to the creation of a habitat suitable for plants to grow.

Author Contributions: Ł.P., A.M.Ł., and R.J.B. designed the study and carried out the field experiments. Ł.P. conducted the majority of the data analysis and wrote the majority of the paper. A.M.Ł. and Ł.P. carried out the biomechanical tests and granulometry. A.M.Ł. described and analyzed the biomechanical test results and prepared bed morphology maps. R.J.B. prepared and depicted bed elevation data. All authors contributed to the final version of the manuscript.

Funding: The research was funded by the National Science Centre, Poland, Grant No. UMO-2014/13/D/ST10/01123 'Field experimental investigation of hydrodynamics of water flow-vegetation-sediment interactions at the scale of individual aquatic plants'.

Acknowledgments: The publication has been partially financed from the funds of the Leading National Research Centre (KNOW) received by the Centre for Polar Studies for the period 2014–2018. We sincerely thank our colleague, Mikołaj Karpiński, for all the help and comments with regard to preparing and conducting field measurements.

Conflicts of Interest: The authors declare no conflicts of interest.

References

1. Aberle, J.; Järvelä, J. Hydrodynamics of vegetated channels. In *Rivers—Physical, Fluvial and Environmental Processes*; Rowiński, P., Radecki-Pawlik, A., Eds.; Springer International Publishing: Cham, Switzerland, 2015; pp. 519–541. ISBN 978-3-319-17718-2.

2. Ormerod, S.J.; Rundle, S.D.; Lloyd, E.C.; Douglas, A.A. The influence of riparian management on the habitat structure and macroinvertebrate communities of upland streams draining plantation forests. *J. Appl. Ecol.* **1993**, 13–24. [CrossRef]

3. McKenney, R.; Jacobson, R.B.; Wertheimer, R.C. Woody vegetation and channel morphogenesis in low-gradient, gravel-bed streams in the Ozark Plateaus, Missouri and Arkansas. *Geomorphology* **1995**, *13*, 175–198. [CrossRef]

4. Simon, A.; Collison, A.J. Quantifying the mechanical and hydrologic effects of riparian vegetation on streambank stability. *Earth Surf. Process. Landf.* **2002**, *27*, 527–546. [CrossRef]

5. Nikora, V. Hydrodynamics of aquatic ecosystems: An interface between ecology, biomechanics and environmental fluid mechanics. *River Res. Appl.* **2010**, *26*, 367–384. [CrossRef]

6. Folkard, A.M. Vegetated flows in their environmental context: A review. *Proc. Inst. Civ. Eng. Eng. Comput. Mech.* **2011**, *164*, 3–24. [CrossRef]

7. Nepf, H.M. Hydrodynamics of vegetated channels. *J. Hydraul. Res.* **2012**, *50*, 262–279. [CrossRef]

8. Yager, E.M.; Schmeeckle, M.W. The influence of vegetation on turbulence and bed load transport. *J. Geophys. Res. Earth Surf.* **2013**, *118*, 1585–1601. [CrossRef]

9. O'Hare, M.T.; Mountford, J.O.; Maroto, J.; Gunn, I.D.M. Plant traits relevant to fluvial geomorphology and hydrological interactions. *River Res. Appl.* **2016**, *32*, 179–189. [CrossRef]

10. Reid, M.A.; Thoms, M.C. Surface flow types, near-bed hydraulics and the distribution of stream macroinvertebrates. *Biogeosciences* **2008**, *5*, 1175–1204. [CrossRef]

11. Osterkamp, W.R.; Hupp, C.R.; Stoffel, M. The interactions between vegetation and erosion: New directions for research at the interface of ecology and geomorphology. *Earth Surf. Process. Landf.* **2012**, *37*, 23–36. [CrossRef]

12. James, C.S.; Jordanova, A.A.; Nicolson, C.R. Flume experiments and modelling of flow-sediment-vegetation interactions. In *Structure, Function and Management Implications of Fluvial Sedimentary Systems; Proceedings of the Symposium on the Structure, Function and Management Implications of Fluvial Sedimentary Systems, Alice Springs, Australia, 2–6 September 2002*; Dyer, F.J., Thoms, M.C., Olley, J.M., Eds.; International Association of Hydrological Sciences, Publication, Institute of Hydrology: Wallingford, UK, 2002; Volume 276, pp. 3–9.

13. Rominger, J.T.; Lightbody, A.F.; Nepf, H.M. Effects of added vegetation on sand bar stability and stream hydrodynamics. *J. Hydraul. Eng.* **2010**, *136*, 994–1002. [CrossRef]

14. Liu, C.; Hu, Z.; Lei, J.; Nepf, H. Vortex Structure and Sediment Deposition in the Wake behind a Finite Patch of Model Submerged Vegetation. *J. Hydraul. Eng.* **2017**, *144*, 04017065. [CrossRef]

15. Bouma, T.J.; van Duren, L.A.; Temmerman, S.; Claverie, T.; Blanco-Garcia, A.; Ysebaert, T.; Herman, P.M.J. Spatial flow and sedimentation patterns within patches of epibenthic structures: Combining field, flume and modelling experiments. *Cont. Shelf Res.* **2007**, *27*, 1020–1045. [CrossRef]

16. Schnauder, I.; Sukhodolov, A.N. Flow in a tightly curving meander bend: Effects of seasonal changes in aquatic macrophyte cover. *Earth Surf. Process. Landf.* **2012**, *37*, 1142–1157. [CrossRef]
17. Cavedon, V. Effects of Rigid Stems on Sediment Transport. Ph.D. Dissertation, University of Trento, Trento, Italy, 2012.
18. Cassan, L.; Belaud, G.; Baume, J.P.; Dejean, C.; Moulin, F. Velocity profiles in a real vegetated channel. *Environ. Fluid Mech.* **2015**, *15*, 1263–1279. [CrossRef]
19. Gurnell, A.M.; Grabowski, R.C. Vegetation–Hydrogeomorphology Interactions in a Low-Energy, Human-Impacted River. *River Res. Appl.* **2016**, *32*, 202–215. [CrossRef]
20. O'Hare, J.M.; O'Hare, M.T.; Gurnell, A.M.; Scarlett, P.M.; Liffen, T.; McDonald, C. Influence of an ecosystem engineer, the emergent macrophyte Sparganium erectum, on seed trapping in lowland rivers and consequences for landform colonisation. *Freshw. Biol.* **2012**, *57*, 104–115. [CrossRef]
21. Sand-Jensen, K. Drag forces on common plant species in temperate streams: Consequences of morphology, velocity and biomass. *Hydrobiologia* **2008**, *610*, 307–319. [CrossRef]
22. Ghisalberti, M.; Nepf, H. Shallow flows over a permeable medium: The hydrodynamics of submerged aquatic canopies. *Transp. Porous Media* **2009**, *78*, 309–326. [CrossRef]
23. Nikora, N.; Nikora, V.; O'Donoghue, T. Velocity profiles in vegetated open-channel flows: Combined effects of multiple mechanisms. *J. Hydraul. Eng.* **2013**, *139*, 1021–1032. [CrossRef]
24. Naden, P.; Rameshwaran, P.; Mountford, O.; Robertson, C. The influence of macrophyte growth, typical of eutrophic conditions, on river flow velocities and turbulence production. *Hydrol. Process.* **2006**, *20*, 3915–3938. [CrossRef]
25. Sukhodolov, A.N. Field-based research in fluvial hydraulics: Potential, paradigms and challenges. *J. Hydraul. Res.* **2015**, *53*, 1–19. [CrossRef]
26. Nikora, V.; Goring, D. Flow turbulence over fixed and weakly mobile gravel beds. *J. Hydraul. Eng.* **2000**, *126*, 679–690. [CrossRef]
27. Sukhodolov, A.N.; Fedele, J.J.; Rhoads, B.L. Structure of flow over alluvial bedforms: An experiment on linking field and laboratory methods. *Earth Surf. Process. Landf.* **2006**, *31*, 1292–1310. [CrossRef]
28. Chanson, H. Acoustic Doppler velocimetry (ADV) in the field and in laboratory: Practical experiences. In Proceedings of the International Meeting on Measurements and Hydraulics of Sewers IMMHS'08, Summer School GEMCEA/LCPC, Bouguenais, France, 19–21 August 2008; Larrarte, F., Chanson, H., Eds.; Department of Civil Engineering, The University of Queensland: Brisban, Australia, 2008; pp. 49–66.
29. Yagci, O.; Kabdasli, M.S. The impact of single natural vegetation elements on flow characteristics. *Hydrol. Process.* **2008**, *22*, 4310–4321. [CrossRef]
30. Thomas, R.E.; McLelland, S.J. The impact of macroalgae on mean and turbulent flow fields. *J. Hydrodyn.* **2015**, *27*, 427–435. [CrossRef]
31. Przyborowski, Ł.; Łoboda, A.M.; Karpiński, M.; Bialik, R.J. Characteristics of flow around aquatic plants in natural conditions: Experimental setup, challenges and difficulties. In *Free Surface Flows and Transport Processes*; Kalinowska, M.B., Mrokowska, M.M., Rowiński, P.M., Eds.; Springer International Publishing: Cham, Switzerland, 2018; pp. 347–361. ISBN 978-3-319-70914-7.
32. Thomas, R.E.; Schindfessel, L.; McLelland, S.J.; Creëlle, S.; De Mulder, T. Bias in mean velocities and noise in variances and covariances measured using a multistatic acoustic profiler: The Nortek Vectrino Profiler. *Meas. Sci. Technol.* **2017**, *28*. [CrossRef]
33. Nikora, V.I.; Sukhodolov, A.N.; Rowiński, P.M. Statistical sand wave dynamics in one-directional water flows. *J. Fluid Mech.* **1997**, *351*, 17–39. [CrossRef]
34. Coleman, S.E.; Nikora, V.I. Fluvial dunes: Initiation, characterization, flow structure. *Earth Surf. Process. Landf.* **2011**, *36*, 39–57. [CrossRef]
35. Kłosowski, S.; Kłosowski, G. *Aquatic and Marsh Plants*; MULTICO: Warsaw, Poland, 2007; ISBN 9788377633557. (In Polish)
36. Łoboda, A.M.; Przyborowski, Ł.; Karpiński, M.; Bialik, R.J.; Nikora, V.I. Biomechanical properties of aquatic plants: The effect of test conditions. *Limnol. Oceanogr. Methods.* **2018**, *16*, 222–236. [CrossRef]
37. Niklas, K.J. Plant Biomechanics. In *An Engineering Approach to Plant Form and Function*; University of Chicago Press: Chicago, IL, USA, 1992; ISBN 0-226-58641-6.
38. ASTM D790-03. *Standard Test Methods for Flexural Properties of Unreinforced and Reinforced Plastics and Electrical Insulating Materials*; ASTM International: West Conshohocken, PA, USA, 2003.

39. MacVicar, B.; Dilling, S.; Lacey, J. Multi-instrument turbulence toolbox (MITT): Open-source MATLAB algorithms for the analysis of high-frequency flow velocity time series datasets. *Comput. Geosci.* **2014**, *73*, 88–98. [CrossRef]

40. Brand, A.; Noss, C.; Dinkiel, C.; Holzner, M. High-resolution measurements of turbulent flow close to the sediment-water interface using bistatic acoustic profiler. *J. Atmos. Ocean. Technol.* **2016**, *33*, 769–788. [CrossRef]

41. Koca, K.; Noss, C.; Anlanger, C.; Brand, A.; Lorke, A. Performance of the Vectrino Profiler at the sediment-water interface. *J. Hydraul. Res.* **2017**, *55*, 573–581. [CrossRef]

42. Goring, D.G.; Nikora, V.I. Despiking Acoustic Doppler Velocimeter data. *J. Hydraul. Eng.* **2002**, *128*, 117–126. [CrossRef]

43. Wahl, T.L. Discussion of 'Despiking acoustic Doppler velocimeter data' by Derek G. Goring and Vladimir I. Nikora. *J. Hydraul. Eng.* **2003**, *126*, 484–487. [CrossRef]

44. Parsheh, M.; Sotiropoulos, F.; Porte-Agel, F. Estimation of Power Spectra of Acoustic-Doppler Velocimetry Data Contaminated with Intermittent Spikes. *J. Hydraul. Eng. ASCE* **2010**, *136*, 368–378. [CrossRef]

45. Voulgaris, G.; Trowbridge, J.H. Evaluation of the acoustic Doppler velocimeter (ADV) for turbulence measurements. *J. Atmos. Ocean. Tech.* **1998**, *15*, 272–289. [CrossRef]

46. Hurther, D.; Lemmin, U. A correction method for turbulence measurements with a 3D acoustic Doppler velocity profiler. *J. Atmos. Ocean. Tech.* **2001**, *18*, 446–458. [CrossRef]

47. Welch, P.D. The use of fast Fourier transform for the estimation of power spectra: A method based on time averaging over short, modified periodograms. *IEEE Trans. Acoust. Electr.* **1967**, *15*, 70–73. [CrossRef]

48. Tuijnder, A.P.; Ribberink, J.S.; Hulscher, S.J. An experimental study into the geometry of supply-limited dunes. *Sedimentology* **2009**, *56*, 1713–1727. [CrossRef]

49. Bialik, R.; Karpiński, M.; Rajwa, A.; Luks, B.; Rowiński, P. Bedform characteristics in natural and regulated channels: A comparative field study on the Wilga River, Poland. *Acta Geophys.* **2014**, *62*, 1413–1434. [CrossRef]

50. Nezu, I.; Nakagawa, H. *Turbulence in Open-Channel Flows*, 1st ed.; CRC Press: Boca Raton, FL, USA, 1993; ISBN 9054101180.

51. Nikora, V. Origin of the "−1" spectral law in wall-bounded turbulence. *Phys. Rev. Lett.* **1999**, *83*, 734. [CrossRef]

52. Gurnell, A.; Goodson, J.; Thompson, K.; Clifford, N.; Armitage, P. The river-bed: A dynamic store for plant propagules? *Earth Surf. Process. Landf.* **2007**, *32*, 1257–1272. [CrossRef]

53. Chambers, P.A.; Prepas, E.E.; Hamilton, H.R.; Bothwell, M.L. Current velocity and its effect on aquatic macrophytes in flowing waters. *Ecol. Appl.* **1991**, *1*, 249–257. [CrossRef] [PubMed]

54. Kemp, J.L.; Harper, D.M.; Crosa, G.A. The habitat-scale ecohydraulics of rivers. *Ecol. Eng.* **2000**, *16*, 17–29. [CrossRef]

55. Carling, P.A. An appraisal of the velocity-reversal hypothesis for stable pool-riffle sequences in the River Severn, England. *Earth Surf. Process. Landf.* **1991**, *16*, 19–31. [CrossRef]

56. Wilkinson, S.N.; Keller, R.J.; Rutherfurd, I.D. Phase-shifts in shear stress as an explanation for the maintenance of pool–riffle sequences. *Earth Surf. Process. Landf.* **2004**, *29*, 737–753. [CrossRef]

57. Tinkler, K.J. Pools, riffles, and meanders. *Geol. Soc. Am. Bull.* **1970**, *81*, 547–552. [CrossRef]

58. Church, M. Bed material transport and the morphology of alluvial river channels. *Annu. Rev. Earth Planet. Sci.* **2006**, *34*, 325–354. [CrossRef]

59. Najafabadi, E.F.; Afzalimehr, H.; Rowiński, P.M. Flow structure through a fluvial pool-riffle sequence–Case study. *J. Hydro-Environ. Res.* **2018**, *19*, 1–15. [CrossRef]

60. Schneider, R.L.; Sharitz, R.R. Hydrochory and Regeneration in A Bald Cypress-Water Tupelo Swamp Forest. *Ecology* **1998**, *69*, 1055–1063. [CrossRef]

61. Västilä, K.; Järvelä, J. Characterizing natural riparian vegetation for modeling of flow and suspended sediment transport. *J. Soils Sediments* **2017**, *17*, 1–17. [CrossRef]

62. Liu, D.; Valyrakis, M.; Williams, R. Flow Hydrodynamics across Open Channel Flows with Riparian Zones: Implications for Riverbank Stability. *Water* **2018**, *9*, 720. [CrossRef]

63. Green, J.C. Modelling flow resistance in vegetated streams: Review and development of new theory. *Hydrol. Process.* **2005**, *19*, 1245–1259. [CrossRef]

64. Aberle, J.; Coleman, S.; Nikora, V. Bed load transport by bed form migration. *Acta Geophys.* **2012**, *60*, 1720–1743. [CrossRef]

65. Venditti, J.G.; Bauer, B.O. Turbulent flow over a dune: Green River, Colorado. *Earth Surf. Process. Landf.* **2005**, *30*, 289–304. [CrossRef]

66. Kostaschuk, R. A field study of turbulence and sediment dynamics over subaqueous dunes with flow separation. *Sedimentology* **2000**, *47*, 519–531. [CrossRef]

67. Singh, A.; Porté-Agel, F.; Foufoula-Georgiou, E. On the influence of gravel bed dynamics on velocity power spectra. *Water Resour. Res.* **2010**, *46*. [CrossRef]

68. Łoboda, A.M.; Bialik, R.J.; Karpiński, M.; Przyborowski, Ł. Two simultaneously occurring Potamogeton species: Similarities and differences in seasonal changes of biomechanical properties. *Pol. J. Environ. Stud.* **2019**, *28*, 1–17. [CrossRef]

69. Puijalon, S.; Bouma, T.J.; Douady, C.J.; van Groenendael, J.; Anten, N.P.; Martel, E.; Bornette, G. Plant resistance to mechanical stress: Evidence of an avoidance–tolerance trade-off. *New Phytol.* **2011**, *191*, 1141–1149. [CrossRef] [PubMed]

70. Biggs, H.; Nikora, V.N.; Papadopoulos, K.; Vettori, D.; Gibbins, C.; Kucher, M. Flow-vegetation interactions: A field study of ranunculus penicillatus at the large patch scale. In Proceedings of the 11th International Symposium on Ecohydraulics, Melbourne, Australia, 7–12 February 2016.

71. Nepf, H.M. Flow and transport in regions with aquatic vegetation. *Annu. Rev. Fluid Mech.* **2012**, *44*, 123–142. [CrossRef]

72. Siniscalchi, F.; Nikora, V. Dynamic reconfiguration of aquatic plants and its interrelations with upstream turbulence and drag forces. *J. Hydraul. Res.* **2013**, *51*, 46–55. [CrossRef]

73. Chen, Y.C.; Kao, S.P. Velocity distribution in open channels with submerged aquatic plant. *Hydrol. Process.* **2011**, *25*, 2009–2017. [CrossRef]

74. Sukhodolova, T.A.; Sukhodolov, A.N. Vegetated mixing layer around a finite-size patch of submerged plants: 1. Theory and field experiments. *Water Resour. Res.* **2012**, *48*. [CrossRef]

MDPI
St. Alban-Anlage 66
4052 Basel
Switzerland
Tel. +41 61 683 77 34
Fax +41 61 302 89 18
www.mdpi.com

Water Editorial Office
E-mail: water@mdpi.com
www.mdpi.com/journal/water

www.ingramcontent.com/pod-product-compliance
Lightning Source LLC
Chambersburg PA
CBHW051907210326
41597CB00033B/6055